A color atlas of

THE RAT

– dissection guide

Published in the U.S.A. and
Canada by Halsted Press, a
Division of John Wiley & Sons, Inc.,
New York
Printed in Italy by Staderini S.p.A.
Library of Congress Catalog Card Number: 78-24742
International Standard Book Number: 0-470-26647-3

A HALSTED PRESS BOOK

A COLOR ATLAS OF
THE RAT

– DISSECTION GUIDE

RONALD J. OLDS,
Ph.D., B.V.Sc., Dp.Bact.

Lecturer, Department of Pathology,
University of Cambridge

and

JOAN R. OLDS, B.Sc.

JOHN WILEY & SONS
New York – Toronto

619.93
OL/c
128164
apn.1984

To our children Jacqueline and Geoffrey
and to all their generàtion

ACKNOWLEDGMENTS

The photograph for 55 was generously provided by Mr J. A. F. Fozzard. We are indebted to the Ministry of Agriculture, Fisheries and Food for permission to publish 58, and to Jane Burton/Bruce Coleman for 59.

We are particularly indebted to T. Gerrard & Co. for generous permission to photograph some of their superb dissections; the results are presented in 15, 21, 26, 37, 38, 39, 41, 42, 49, 50 and 57. The staff of the Animal Unit of the Department of Pathology, University of Cambridge have been most helpful in finding suitable rats for the remaining specimens, including all the fresh preparations.

We have been encouraged and helped by critical comments on the text by Mr K. D. Taylor and Mr W. Lane-Petter. The publishers have been very helpful and patient.

It is a pleasure to acknowledge the assistance of all of these persons and to record our thanks for their contributions.

Contents

Preface

This atlas is intended to meet the needs of animal technicians and of secondary school and university students in their first studies of rat anatomy. Many students have only a brief view of an animal dissected for a group; some may never see more than one rat dissected. The photographs in this book provide a permanent record of the appearance of the tissues, usually in the fresh state.

If dissection specimens are not freely available, it may be helpful for the student to study the book before he sees the dissection, so that he is well prepared. A photographic atlas will not render the dissection unnecessary, but it should reduce wastage of the available animals, and aid the student's recall.

Each photograph is accompanied by essential explanatory text describing how the animal was prepared for photography, with a general review of the main features illustrated. The particular tissues and organs are identified by the key to the numbers superimposed on the photographs. The relevant dissection procedure to be followed is given for each photograph.

The last section of the book is a description of the general biology of the rat, with a table showing the main differences between *Rattus norvegicus* and *Rattus rattus*.

The authors believe that the simplicity of the text represents the depth of treatment required by the great majority of students of rat structure and function, and they have avoided burdening the majority with anatomical detail which is of interest only to a minority. More detailed descriptions can be found in Greene (1935), Smith (1968) and Zeman and Innes (1963).

Standard methods were used for photography. For dissection, a black background was frequently used to contrast with the tissues. Some of the dissections were illuminated from the caudal end: this proved to be the most satisfactory method for preventing the smaller abdominal organs being obscured by the shadow of the liver. The photographs have been printed with the head uppermost; the lighting, which thus appears to come from below, may seem slightly unnatural. By examining the position of the shadows or of the highlights, however, the viewer may easily determine the direction of the illumination.

Words of position in this atlas always refer to the animal in its normal standing attitude.

For example, 'left' refers to the left side of the rat and not, necessarily, to the left side of the reader. Moreover, because 'dorsal', for example, refers to the top surface of a standing quadruped, some terms will differ from the corresponding terms of anatomy of bipedal man.

1 An albino laboratory rat This adult male was placed in a clear plastic cage for photography. It spent the next half hour actively investigating its new environment. It is a male Wistar, a laboratory albino strain of the brown rat, *Rattus norvegicus*.

Even when the animal raises itself on its haunches, its back still retains the characteristic hump. In this position its fore limbs are free to handle food and to groom. Note the guard hairs on the back and the vibrissae.

Dissection procedure
If possible examine the live animal preferably over a period of time in the laboratory, and note the general features of its behaviour, its feeding, climbing, exploration, mating and the use of its limbs.

1 rhinarium
2 skin folded into the
  diastema
3 incisors
4 mystacial vibrissae

2 The incisor teeth and the diastema The jaws of a dead rat were opened slightly for photography.

The rat has one incisor and three molars on each side of the upper and lower jaw (see **31**). Thus, its dental formula is $I_1^1 \, C_0^0 \, PM_0^0 \, M_3^3 = 16$. As in all rodents the incisors grow continuously and are honed by wear into sharp chisels specialised for gnawing. The resulting debris from gnawing inedible material need not be swallowed, since folds of skin behind the incisors can seal off the back of the oral cavity. The space between the incisor and molar teeth is called the diastema.

The nostrils are at the edge of a hairless zone called the rhinarium; they can be closed under water.

Dissection procedure
In the freshly-killed animal, look in greater detail at the external features, ears, eyes, nostrils, mouth, vibrissae.

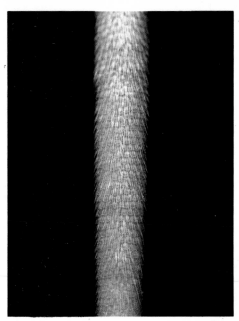

3 The ventral surface of the tail The entire surface of the tail is covered by rows of scales which overlap like roof-tiles. Three short bristles project from under the edge of each scale. The surface is covered with orange-yellow, waxy grease.

Dissection procedure
Examine the scaly surface of the tail, and note the relationship of the hairs to the epidermal scales.

1 claw
2 digital pad
3 footpad
4 first digit
5 pre-axial border
6 post-axial border
7 tarsal joint

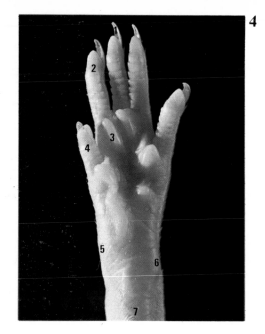

4 The ventral surface of the left hind foot All five digits are well developed. The hairless surface extends to the tarsal joint. There are six footpads on the sole. The hind limbs are larger than the fore.

The rat's gait is plantigrade, as shown by the hairless sole extending from the toes to the tarsal joint. This joint corresponds to the ankle of man or the hock of the horse, for example.

The considerable difference in the size of the fore and hind limbs causes the rat to run at speed with irregular darting-hopping movements which help it to avoid capture. (This action is more marked in some other rodents such as the squirrel, and even more so in the jerboa which has a bipedal hopping gait.)

Dissection procedure
Look in detail at the hind limb.

1 claw
2 digital pad
3 footpad
4 reduced first digit
  (thumb or pollex)
5 pre-axial border
6 post-axial border

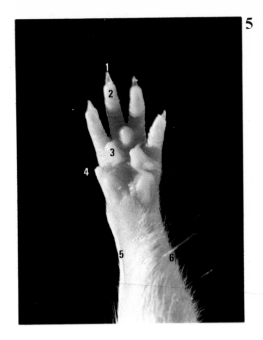

5 The ventral surface of the right fore foot The first digit (the thumb or pollex) is reduced to a small stub, the flattened nail of which can be felt more readily than it can be seen. The four remaining well-developed digits are therefore numbered II to V from the pre-axial border; each of these terminates in a claw shaped like a curved pyramid, which is hollow from the ventral surface. The hairless sole of the foot has a number of pads: a digital pad at the apex of each digit, and five larger footpads on the palm.

Dissection procedure
Look in detail at the fore limb.

1 thoracic nipples
2 axillary nipples
3 abdominal nipples
4 inguinal nipples
5 clitoris

6 Abdomen of a mature female, ventral view This female was killed immediately after her litter was weaned at four weeks' old. The teats are especially obvious, since they are near their maximal development, and the hair around them has been flattened by the youngsters' suckling. In the non-pregnant, non-lactating female the nipples are often hidden by hair.

There are usually six pairs of teats: three thoracic, one abdominal and two inguinal. Sometimes, as in this specimen, one of the second thoracic teats is lacking.

Dissection procedure
In the female, determine the distribution of the nipples; in a lactating animal, they will be readily visible, but in a non-lactating female they are best located by rubbing the hand over the ventral surface of the abdomen.

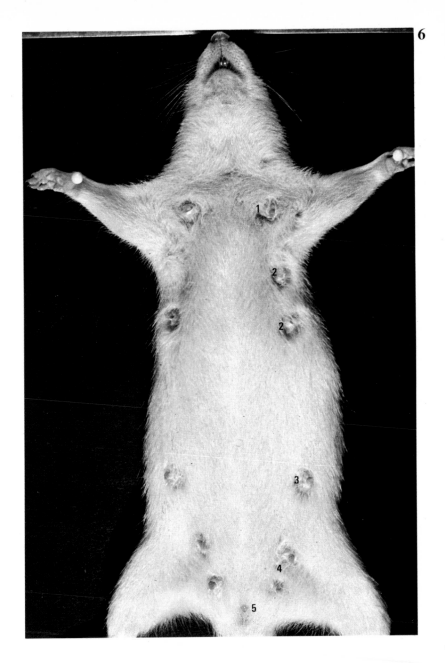

1 aperture of the prepuce
2 scrotal sacs
3 tail

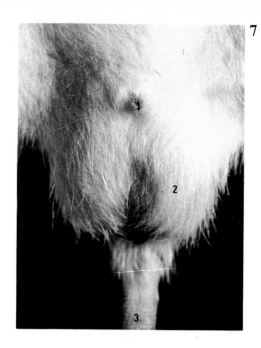

7 External genitalia of the male A mature animal photographed from the ventral surface.

There is a common urinogenital aperture in the male rat; compare this with the female (see **8**). When viewed through the intact skin, the scrotal sacs appear to be fused in front and partially separated behind. Later it will be seen that the scrotal sacs are quite separate internally (see **18**). The large testicles produce a rather protuberant scrotum. This gives the posterior line of the animal's trunk a rounded or cylindrical shape, viewed from either the ventral or dorsal aspects. The anus is obscured by the projecting scrotum and its hair.

The testes are situated in the abdomen in the newborn male (see **10**). They do not descend into the scrotum until the animal is four to six weeks' old. The inguinal canal remains open in the adult rat, so that the testes may be withdrawn into the abdomen.

Dissection procedure
Note the visible features of the external genitalia of the male.

1 clitoris with urinary
  aperture
2 vulva, with the genital
  aperture
3 anus
4 base of the tail

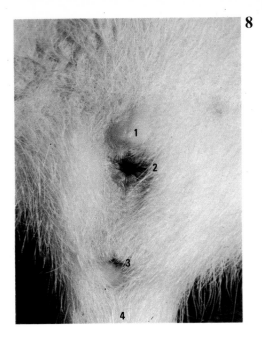

8 External genitalia of the female The urethra opens at the base of the clitoris, a small projecting homologue of the penis, which is enclosed in a little prepuce. Between the clitoris and the anus, and nearer the former, is the vagina which is open in this animal; it would be closed by a membrane during the first ten weeks of life, that is, until puberty.

Note that the rump of the female tapers to the tail in a characteristic triangular shape (cf. **7**). The sex of a mature animal can be determined by inspection from above, simply by determining the shape of the posterior line of the trunk: triangular in the female, rounded in the male.

Dissection procedure
Note the visible features of the external genitalia of the female.

1 prepuce inverted
2 penis
3 scrotal sac

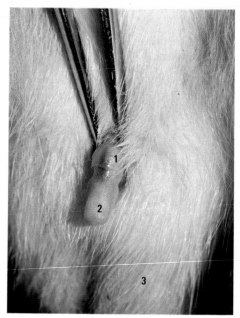

9 The penis displayed *in situ* The penis has been extruded by gentle backwards pressure at the sides of the prepuce, and then held with forceps for photography.

The penis is protected by a loose sheath called the prepuce. In the floor of the penis there is a bony plate, the *os penis* (see **56**), which projects beyond the tip of the penis. The external orifice of the urethra is the aperture common to the urinary and genital tracts in the male. It is a small slit at the tip of the penis.

During copulation the erection of the penis causes its extrusion from the prepuce. Erection results from the engorgement of dense vascular tissue with blood.

Relatively few animal species have an *os penis*. It is found in other rodents but, of the common domestic animals, only in the dog.

Dissection procedure
Extrude the penis and examine it.

10

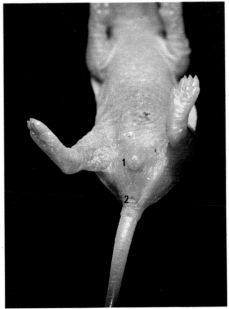 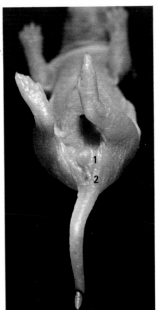

1 genital papilla
2 anus

10, 11 Determination of sex of newborn rat These two littermates died
before they were 24 hours old. They were supported on dissection pins
for photography, and the tail held back with black cotton thread.

It is often useful to know how to determine the sex of the newly-born
animal. At this stage, the genital papilla of the male is rather larger than
that of the female, but not markedly so. A more reliable guide is the
distance between the genital papilla and the anus. It is greater in the male
(**10**) than in the female (**11**); the mean values given by Farris and
Griffiths (1962) are 2.8mm for the male, and 1.2mm for the female.

The newborn animal is hairless, and its ears and eyes are closed. The
ears open at about three days, and the eyes at 14–17 days. Some of the
abdominal viscera can be seen through the thin skin of the newborn rat.

19

12 Subcutaneous tissues of a female The first stage of the dissection is just completed.

Note that the mammary tissue in this animal is quite inconspicuous (see **13**). The structures in the throat will be examined more fully in **14**.

The white line (*linea alba*) is a band of white fibrous tissue running in the mid-ventral line of the abdomen. It is the junction between the abdominal muscles, chiefly the oblique and transverse muscles. Through the relatively thin abdominal musculature the abdominal viscera can be seen (see **18**). The most obvious of these are the liver, stomach, caecum and small intestine.

Dissection procedure
Pin the carcass on a dissecting board; insert the pins through the palms and soles of the feet or, for the hind limbs, between the hamstrings and the tibia. Make an incision through the skin only, from the chin to the root of the tail, that is, the full length of the ventral surface: cut to one side of the genitals. Then cut at right angles to this along the fore limbs as far as the elbows. Make similar incisions along the hind limbs to about the level of the knee joint. Reflect the skin with the fingers, or by blunt dissection with blunt-pointed scissors; in so doing, avoid damaging the superficial nerves and blood vessels. Pin back the flaps of skin.

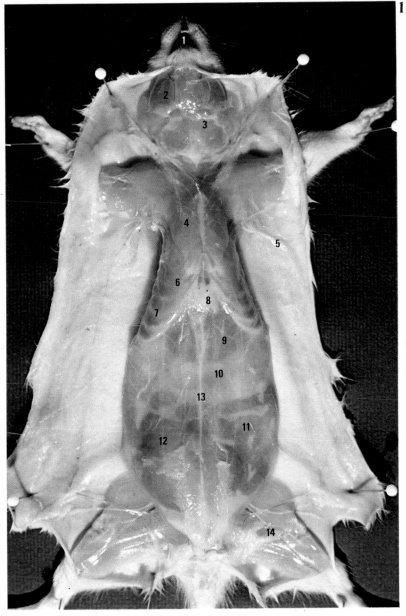

1 mouth
2 masseter muscle
3 thoracic nipples
4 axillary nipples
5 anterior mass of
   mammary tissue
6 posterior mass of
   mammary tissue
7 abdominal nipple
8 inguinal nipple
9 preputial gland

13 Subcutaneous tissues of a lactating female This is the animal shown in **6**, which has just weaned its litter. The skin was reflected to expose the mammary tissue. No attempt was made to separate the glands from the *panniculus carnosus* muscle to which they are intimately attached. The teats have been left attached to the underlying glandular tissues, and the skin cut round them on the left side.

The very extensive mammary tissue clearly consists of anterior and posterior portions. The anterior mass extends as far as the chin, and up the sides of the neck and thorax. The posterior part is separated from the anterior by a space just behind the ribs. The tissues of the abdominal and inguinal glands are confluent. The mammary tissue shown is red-brown in colour but, if it were full of milk, it would be paler and mottled with white.

Dissection procedure
If the animal is a female, determine the distribution of the mammary tissue.
In the non-lactating animal the tissue will not be so obvious as that shown in **13**. It might be necessary to remove some mammary tissue before continuing with the dissection of the throat region.

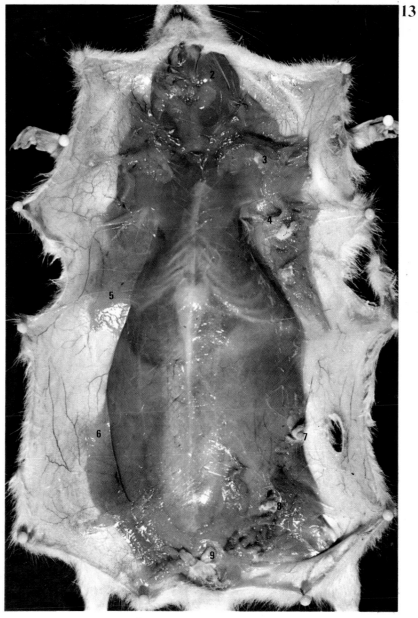

1 upper incisors
2 lower incisors
3 masseter muscle
4 parotid duct
5 exorbital lachrymal
   gland
6 lymph nodes
7 sublingual gland
8 parotid gland
9 submaxillary gland
10 external jugular vein
11 pectoral muscle

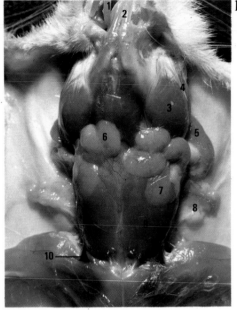

14 Superficial dissection of the ventral aspect of the neck The skin was reflected as described for **12**.

Of the salivary glands, the most obvious from this view are the submaxillary glands, which are in contact along the mid-ventral line, and cover most of the ventral surface of the neck. The major sublingual gland is closely applied to the anterior-lateral surface of the sub-maxillary gland. The parotid is a less compact structure which extends from the ventral surface behind the jaw to the ear. The ducts of these salivary glands enter the buccal cavity independently.

This region is very well supplied with lymph nodes. The submaxillary lymph node is embedded in part of the parotid gland. Anterior to it there are usually two other nodes, but the number and distribution of lymph nodes appears to be variable; they are often not bilaterally symmetrical.

The normal wearing of the teeth is dependent on their meeting properly; an un-apposed tooth may continue to grow until it enters the skull or damages other tissues. This animal had congenitally deformed

Dissection procedure
Examine the salivary glands, lymph nodes and other structures in the throat region.

1 mandible
2 digastric muscle
3 submental artery and vein
4 masseter muscle
5 anterior facial vein
6 posterior facial vein
7 larynx
8 external carotid artery
9 internal carotid artery
10 thyroid
11 trachea
12 common carotid artery
13 external jugular
14 internal jugular

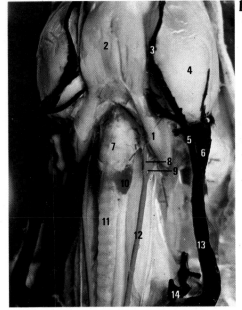

15 The major blood vessels of the neck This is a more detailed view of the specimen prepared as described in **27**. In this injected preparation the salivary glands, lymph nodes and muscles and some minor blood vessels have been removed.

The thyroid is clearly demonstrated because of the plexus of veins on its surface (see **17**).

Dissection procedure
Refer to an injected preparation for assistance in identification of the blood vessels.

***

14 Continued

teeth. Two weeks after weaning it was noticed that it was very thin and it was killed. Because this animal was undernourished, it lacked all signs of adipose tissue and it was, therefore, an excellent dissection subject for some purposes, although abnormal in certain respects. The absence, along with the other adipose tissue, of the multilocular adipose tissue supports the opinion that this so-called 'hibernating gland' is a particular type of adipose tissue.

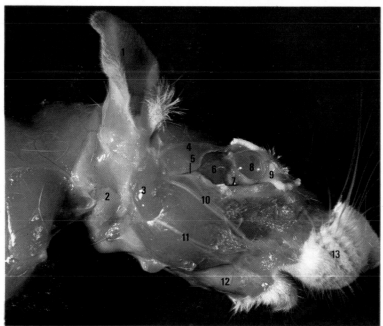

1 pinna	**7** harderian gland
2 parotid salivary gland	**8** eyeball
3 exorbital lachrymal gland	**9** intraorbital fat
4 duct of exorbital lachrymal gland	**10** facial nerve
5 superficial temporal artery and vein	**11** parotid duct
	12 mandible
6 intraorbital lachrymal gland	**13** insertion of vibrissae in rows

16 Lateral view of the superficial tissues of the head and neck The skin was completely removed from this specimen, except for that around the muzzle which is very firmly attached by the structures surrounding the follicles of the mystacial vibrissae, and that over the ears.

Dissection procedure
Reflect the skin off the sides of the face, and examine the structures so revealed.

1 larynx
2 parathyroid
3 thyroid
4 isthmus of thyroid
5 trachea
6 inferior thyroid vein

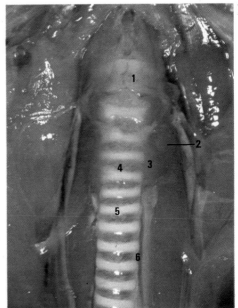

17 Trachea, larynx and thyroid The overlying salivary glands, lymph nodes and muscles were removed down to the ventral surface of the trachea and associated structures.

The trachea consists of a series of cartilaginous rings linked by bands of fibrous tissue and muscle. This produces a flexible tubular structure which is firmly supported to give a clear airway irrespective of the position of the neck. The larynx consists of a number of plates of cartilage which form a box housing the vocal cords, *i.e.* the 'voicebox'.

The thyroid is a pale orange-coloured gland consisting of two lobes, one on either side of the trachea just behind (caudal to) the larynx. They are joined by a delicate isthmus across the ventral surface of the trachea (see **15**).

The rat has two parathyroids, one laterally situated at the anterior pole of each thyroid lobe. They are circular and paler than the thyroid.

Dissection procedure
In the mid-line dissect away the glands and muscles to expose the length of the trachea, the larynx and the thyroid.

1 thorax
2 xiphoid process
3 liver
4 stomach
5 omentum
6 spleen
7 caecum

8 colon
9 small intestine
10 fat
11 urinary bladder
12 prostate gland
13 preputial gland
14 penis
15 scrotal sac

18 Abdominal viscera of a male *in situ* The carcass was chilled (4°C) for a few hours before dissection to set the organs firmly in place, and to reduce haemorrhage.

The testes lie in their scrotal sacs. Part of the small intestine is filled with dark semi-digested fibrous food. The contents of the caecum are paler. There is no sign of post-mortem decomposition.

Dissection procedure
Note that some organs can be seen through the muscular wall of the abdomen (12). This gives an indication of the thickness of the musculature. With forceps pick up this muscle layer, leaving the underlying viscera undisturbed. Cut with scissors through to the peritoneum; this will produce a V-shaped opening about 0.5–1cm long. Insert one tip of the scissors through this gap, and with them make a cut from the xiphoid process of the sternum to the front of the pelvis. Still using the scissors, cut on each side through the muscle wall immediately behind the posterior edge of the ribs, until the flaps of skin so formed can be pinned out as in **18**. Identify the organs *in situ*.

1 xiphoid cartilage
2 liver
3 spleen
4 stomach
5 part of pancreas in gastrosplenic
  omentum
6 duodenum
7 ileum

8 mesentery
9 caecum
10 colon
11 rectum
12 uterus
13 fat
14 urinary bladder
15 vagina
16 preputial gland

19 Alimentary tract displayed within the abdomen The small and large intestines were displaced to the right, the stomach to the left, and the spleen was exposed by pulling it slightly around the stomach. No fat has been removed at this stage.

The stomach is relatively empty. The purple-red colour of the spleen distinguishes it from the liver, which is normally mahogany-brown. The lobes of the liver are better seen in **23** and **33**. The pancreas is diffuse: part lies within the U-shaped bend of the duodenum and from here it ramifies through to the gastro-splenic omentum.

The rectum runs straight back to the anus and contains faecal pellets. It lies dorsal to the internal genitalia.

Dissection procedure
Displace the small and large intestines to the right and identify the major abdominal viscera now displayed.

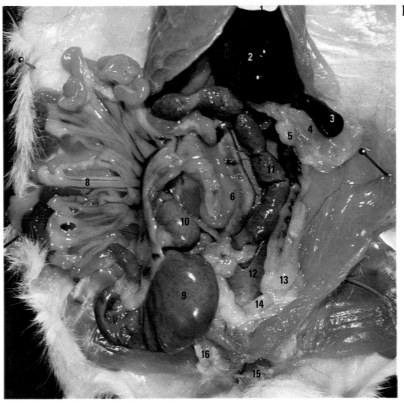

1 costal margin
2 liver
3 stomach
4 spleen
5 left adrenal
6 posterior adrenal vein
7 left kidney
8 renal vein
9 part of the solar plexus
10 lymph nodes
11 omentum
12 rectum
13 small intestine
14 internal spermatic vein
15 ureter and ureteric
   vessels
16 fat

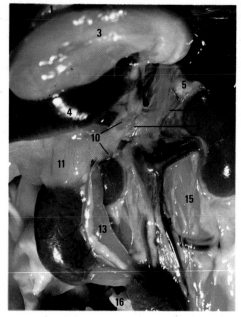

20 Kidneys, adrenals and neighbouring viscera in a male rat. The gut and the stomach with the attached omentum, spleen and pancreas have been displaced to the right to expose the left kidney and the adjoining viscera. The posterior pole of the right kidney is also visible, although in most rats it is further forward than this.

The healthy rat kidney is brown, bean-shaped and has a smooth surface. The adrenal has a pale orange-yellow surface. In general the adrenal is larger in the female and in the wild rat and larger in the black rat than in the brown.

Dissection procedure
Gently displace the stomach, spleen and omentum to the right, and examine the kidneys and adjacent tissues in the lumbar region.

1 liver
2 posterior vena cava
3 abdominal aorta
4 adrenal
5 posterior adrenal vessels
6 coeliac axis
7 anterior mesenteric artery
8 left kidney
9 renal vessels
10 ureter
11 right kidney
12 posterior mesenteric artery and vein
13 left iliolumbar vessels
14 left internal spermatic vessels
15 right internal spermatic vessels
16 right iliolumbar vessels
17 common iliac vessels
18 rectum

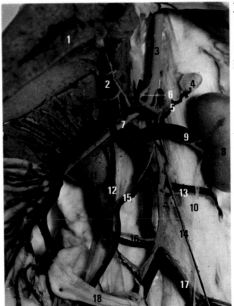

21 The major abdominal blood vessels In this injected preparation the stomach, spleen, small intestine and part of the liver have been removed, and the rest of the liver deflected to the right to expose the vessels of the lumbar region. The mesenteric vessels are displayed on the surface of the remaining part of the liver. Some of the smaller blood vessels have been removed or cut. This is a detailed view of part of the specimen prepared as described in **27**.

Dissection procedure
An injected preparation will assist in the identification of the blood vessels of this area.

1 muscle wall of abdomen
2 rib
3 adrenal
4 kidney
5 aorta
6 coeliac artery
7 coeliac plexus
8 anterior mesenteric artery
9 lymph node
10 small intestine
11 caecum
12 rectum
13 ureter
14 internal spermatic vein
15 scalpel holding kidney aside

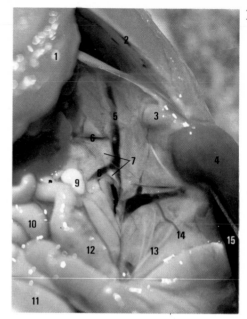

22 The coeliac (solar) nerve plexus A lean carcass was selected, and an incision was made through the abdominal wall on the left side close to the vertebral column. Most of the viscera were displaced to the right side and a small amount of acid-alcohol was poured on the area which turned the nerves and lymph node opaque-white. The kidney was retracted with a scalpel blade for photography.

From the coeliac plexus, numerous fine nerves arise which form secondary plexuses along the neighbouring arteries.

If the animal were not lean, the coeliac plexus would be covered by fat and not so readily seen.

Dissection procedure
The solar plexus may be found at this stage, but it is best demonstrated on a lean carcass using a different dissection approach.

1 xiphoid cartilage
2 falciform ligament
3 heart
4 lung
5 central tendinous part of
  diaphragm
6 muscular part of
  diaphragm
7 phrenic veins
8 liver

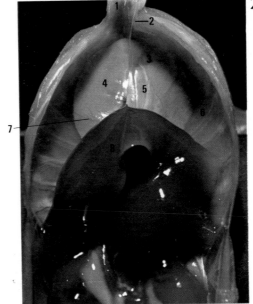

23 The diaphragm viewed from the abdominal side To expose the diaphragm more clearly, the stomach was removed and the animal pinned so that the thorax was horizontal. The rest of the body was flexed over an edge to allow the liver and the other abdominal organs to fall away from the diaphragm.

Since the thorax is intact, the diaphragm has not collapsed away from the lungs. The diaphragm is thin, which allows the ribs, lungs and a portion of the heart to be seen through it.

Dissection procedure
Examine the abdominal surface of the diaphragm. This is easiest if the abdominal viscera can be held out of the way with a blunt instrument – in **23** some organs have been removed and others allowed to drop away under gravity.

1 first rib
2 pectoralis muscle
3 thymus
4 right auricle
5 ventricles
6 right lung
7 left lung
8 phrenic nerves
9 posterior vena cava
10 oesophagus
11 aorta
12 phrenic vein
13 muscular part of
   diaphragm
14 central tendinous part
   of diaphragm

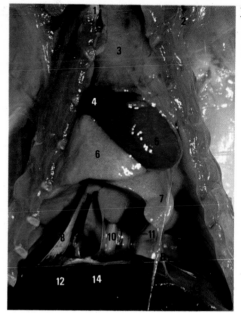

24 Thoracic viscera *in situ* The ventral wall of the thorax has been removed by cutting through the ribs on either side of the sternum.

It will be appreciated in cutting through the ribs that they are calcified; the rat does not have typical costal cartilages.

In this 3 months' old female, the thymus is at its greatest development; the size of the gland decreases with age.

The lungs are contracted away from the thoracic wall; they are elastic and collapse as soon as the thorax is opened – often more so than in this specimen.

Dissection procedure

Cut through the diaphragm near the xiphoid cartilage, and carry this cut with scissors about 2cm either side of the mid-line. Now turn the scissors at right angles to the previous cut, and cut forward through the ribs to remove the floor (ventral wall) of the thorax.

1 muscles and ribs
  reflected ventrally
2 remainder of thoracic
  wall held back with
  black cotton thread
3 lymph nodes
4 left anterior vena cava
5 azygos vein (present
  only on left side)
6 area of confluence of
  left and right
  anterior vena cava
  and posterior vena
  cava
7 dorsal aorta
8 left pulmonary vein
9 left auricle
10 thymus
11 ventricle
12 left lung held back with
  black cotton thread
13 post-caval lobe of right
  lung
14 left phrenic nerve
15 posterior vena cava
16 diaphragm
17 stomach
18 liver

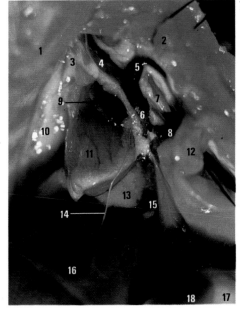

25 The heart and some nearby organs An incision was made through the thoracic wall on the left side close to the vertebral column. Black cotton threads were used to hold apart the two portions of the thoracic wall and to retract the left lung. (This additional dissection illustrates some features that are difficult to see with the usual dissection procedure.)

The left phrenic nerve can be traced across the left anterior vena cava and the heart to the diaphragm. The azygos vein and some of the pulmonary veins can be seen *in situ*. Note that the left anterior vena cava passes behind and beneath the other great vessels on the dorsal surface of the heart to enter the right auricle with the right anterior vena cava and the single, posterior vena cava (see **30**).

Dissection procedure
Examine the thoracic organs *in situ*. Pin the heart aside and identify the major blood vessels around the heart.

1 mandible
2 larynx
3 left common carotid artery
4 external jugular vein
5 right common carotid artery
6 right subclavian artery
7 innominate artery
8 left subclavian artery
9 brachial artery and vein
10 lateral thoracic artery and vein
11 arch of aorta
12 heart
13 left lung
14 lobes of right lung
15 posterior vena cava
16 aorta
17 diaphragm
18 liver
19 mesenteric vessels displayed on the liver and on the right knee
20 left adrenal artery and vein
21 left kidney
22 renal artery and vein
23 anterior mesenteric artery and vein
24 iliolumbar artery and vein
25 posterior mesenteric artery and vein
26 right internal spermatic artery and vein
27 left internal spermatic artery and vein
28 common iliac artery and vein
29 external iliac artery and vein
30 femoral artery and vein
31 pampiniform plexus

26 An injected preparation to show the major blood vessels Blue latex was injected into the venous system; it passed through the right side of the heart and coloured the lungs as well. Red latex was injected into the left ventricle, from which it entered the arteries. In this preparation, part of the abdominal and thoracic walls, a number of muscles and other superficial tissues, the alimentary tract and portions of the liver have been removed to uncover the major arteries and veins.

The lobes of the lungs are displayed. Note that the coronary veins can be seen on the ventricle; the heart has its own blood supply.

Dissection procedure
Refer to an injected preparation to identify the blood vessels.

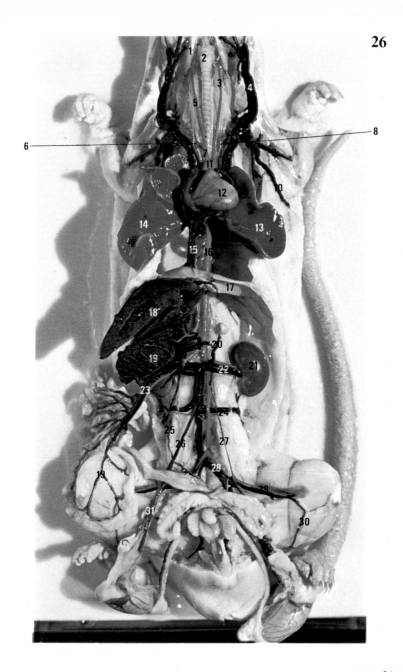

1 larynx
2 trachea
3 common carotid artery
4 brachial plexus
5 axillary artery
6 muscle reflected to
   expose the brachial
   plexus
7 arch of the aorta
8 left anterior vena cava
9 right auricle
10 pin holding the heart
   aside
11 left lung
12 posterior vena cava
13 oesophagus
14 left phrenic nerve
15 dorsal aorta

27 Brachial plexus and axillary blood vessels The muscles of the upper fore limb have been dissected to display the brachial nervous plexus and the blood vessels of the fore limb. The heart has been pinned aside to display some of the major blood vessels.

In this specimen the lungs are quite collapsed. Part of the lung is the normal bright pink but some (for example, the anterior part of the left lobe) has a colour and consistency like that of the liver. This so-called 'hepatised' area is affected by pneumonia – a common disease of rats.

Dissection procedure
Dissect through the muscles beneath the scapula to display the brachial plexus of nerves, and the blood vessels supplying the fore limb.

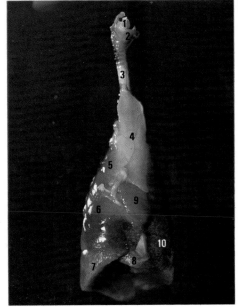

1 epiglottis
2 thyroid cartilage
3 trachea
4 thymus
5–8 lobes of right lung
  5 anterior lobe
  6 median lobe
  7 posterior lobe
  8 post-caval lobe
9 heart
10 left lung

28 The thoracic viscera displayed The thoracic viscera were fixed *in situ* by introducing 10 per cent formalin into the trachea. After 2 days fixation, the thoracic viscera were removed along with the trachea and larynx.

The organs retain the shapes and relationships that they have in the living animal, but the lungs are discoloured brown by the formalin.

Dissection procedure
The relationships of the thoracic organs to each other is best demonstrated on a separate animal by fixing them briefly *in situ*, before removing them from the thorax for examination. (If the organs are removed in the unfixed state, the lungs will collapse. However, if they are handled with care, they may be reinflated through a glass or plastic tube tied into the trachea; this demonstrates their normal pink colour, their elasticity and the shapes of the lobes.)

1 epiglottis
2 thyroid cartilage
3 trachea
4 thymus
5–8 lobes of right lung
  5 anterior lobe
  6 median lobe
  7 posterior lobe
  8 post-caval lobe
9 heart
10 left lung

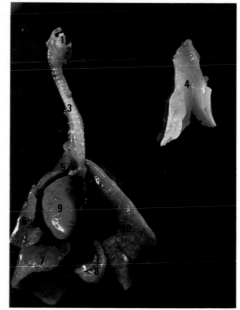

29

29 The thoracic viscera displayed further From the specimen used in **28**, the thymus was removed and placed aside. The lungs were retracted slightly to show the various lobes.

Dissection procedure
Remove the thymus. Separate the remaining thoracic organs and examine them further.

1 aorta
2 left anterior vena cava
3 right anterior vena cava
4 posterior vena cava
5 pulmonary artery
6 pulmonary veins
7 bronchus
8 trachea
9–11 lobes of right lung
   9 median lobe
  10 posterior lobe
  11 post-caval lobe
12 left lung
(13 shadow of aorta)

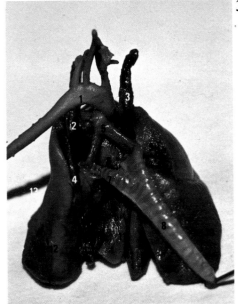

30 The complex of vessels on the dorsal surface of the heart Green latex was injected into the right ventricle and hence into the pulmonary arteries; blue latex was injected into the systemic veins. Red latex was injected into the left ventricle and through the left auricle into the pulmonary veins; because the wall of the aorta is thicker than that of the pulmonary veins, the aorta is coloured light pink and the pulmonary veins a darker brown-red. The oesophagus was dissected off the trachea, which was then folded backwards and pinned in this position. The aorta was pinned to the left; it casts a shadow in this photograph.

The first three branches of the aorta are: the innominate artery which soon divides into the right common carotid and right subclavian arteries; the left common carotid; the left subclavian arteries.

Note that the posterior vena cava and the left anterior vena cava sweep under and behind the other vessels to enter the right auricle.

Dissection procedure
Remove the oesophagus. Turn back the trachea, and examine the arrangement of the major blood vessels entering the dorsal surface of the heart; refer to an injected preparation.

1 rhinarium
2 cleft lip
3 incisors
4 vibrissae
5 folds of skin over diastema
6 rugae of hard palate

7 molars
8 soft palate
9 masseter muscles (cut)
10 condyle of mandible
11 pharynx
12 tongue
13 trachea

31 The opened buccal cavity The skin was reflected, the masseter muscles were cut through on each side, and the lower jaw was turned through about 180° to display the contents of the mouth cavity. Some tissues were torn by this procedure, and the articulating surfaces of the mandibles, the condyles, were dislocated and exposed. The jaws were held open by black cotton thread.

Note the well-developed masseter muscle. This gives another view of the incisors and molars and of the diastema and its skin folds (see **2**).

Dissection procedure
Open the buccal cavity by cutting through the masseters on each side; turn the lower jaw through about 180 degrees and examine the structures of the buccal cavity.

1 incisors
2 tongue
3 molars
4 trachea
5 heart
6 lungs
7 oesophagus
8 stomach
9 liver

10 bile duct
11 spleen
12 pancreas (2 portions)
13 duodenum
14 small intestine
15 caecum
16 appendix
17 colon
18 rectum
19 anus

32 The entire alimentary tract displayed The alimentary canal was removed, as described below.

The oesophagus is closely associated with the larynx and the trachea. The spleen is loosely attached to the greater curvature of the stomach. The characteristic U-shape of the duodenum encloses one part, the 'head', of the pancreas. The remainder of the small intestine is coiled and suspended by mesentery. A constriction about the middle of the caecum divides it into a large basal part and the apex; the latter contains a mass of lymphoid tissue in its walls and corresponds to the appendix. The colon and rectum contain formed pellets of faeces.

Dissection procedure
On a separate animal, remove the lower jaw, taking with it the oesophagus, the larynx, and the remainder of the trachea. Remove the entire alimentary tract together with the liver, spleen, omentum, pancreas and part of the skin around the anus. Arrange it as shown.

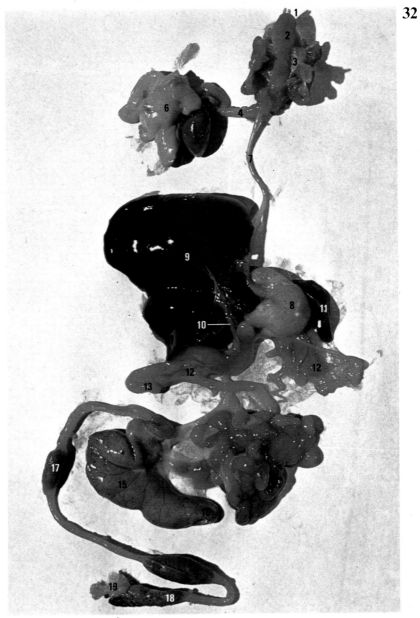

1–4 lobes of liver
   1 left lobe
   2 median lobe
   3 right lobe
   4 caudate (spigelian) lobe
5 oesophagus

6 stomach
7 spleen
8 pylorus
9 bile duct
10 duodenum
11 pancreas

33 The stomach, the liver and associated structures These tissues were arranged to display the relationships of the stomach, liver, duodenum, spleen and pancreas.

The stomach shows externally its division into a cardiac part, which is translucent with a white lining, and a pyloric part which has an opaque thicker wall.

There is no gall bladder in the rat. The bile duct enters the duodenum about 2cm from the pylorus.

Dissection procedure
Re-examine the relationships of the stomach, spleen, liver, pancreas and duodenum.

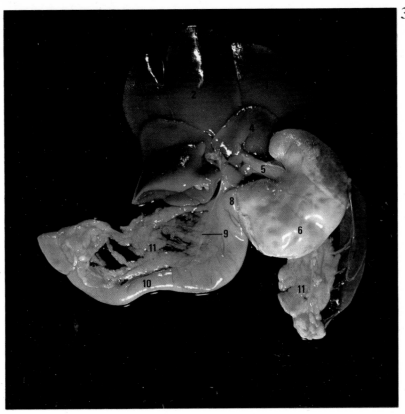

1 part of duodenum
2 jejunum
3 ileum
4 ileocaecal junction
5 mesenteric lymph nodes

6 mesentery containing blood
  vessels and lymphatics
7 caecum
8 apex of caecum
9 colon

34 The tissues at the junction of the small and large intestines The ileo-caecal junction was located, and the terminal part of the small bowel was arranged to display the mesentery and the mesenteric lymph nodes.

The ileo caecal junction represents the division between the small and large intestines.

The caecum is a blind sac; in this specimen there is no external demarcation between its basal and apical portions. The caecum leads directly into the colon in which water is absorbed and residues are inspissated to form faecal pellets.

The mesentery of the jejunum and ileum contains arteries, with their plexuses of nerves, veins, lymphatics (lacteals) and a variable amount of fat.

The lacteals converge on the mesenteric lymph node which is situated at the root of the mesentery and is the largest lymph node in the rat.

The lymph nodes are difficult to demonstrate in a more obese animal since they are embedded in fat.

Dissection procedure
Re-examine the relationships of the structures near the junction of the small and large intestines.

1 incisors	**7** small intestine
2 tongue	**8** ileocaecal junction
3 molars	**9** caecum
4 trachea	**10** appendix
5 oesophagus	**11** colon
6 stomach	**12** rectum
	13 anus

35 The alimentary tract displayed further The gut shown in **32** was further dissected. The respiratory tract was removed. The mesentery and other attachments were severed with scissors, and the gut arranged in straight lengths to show the relative dimensions of the various parts of the alimentary tract.

The large intestine of the rat is about as long as its body (*i.e.* from nose to anus); the small intestine is about six times this length.

Dissection procedure
Sever the mesentery and other attachments with scissors and arrange the gut in straight lengths to show the relative proportions of the various parts of the alimentary tract.

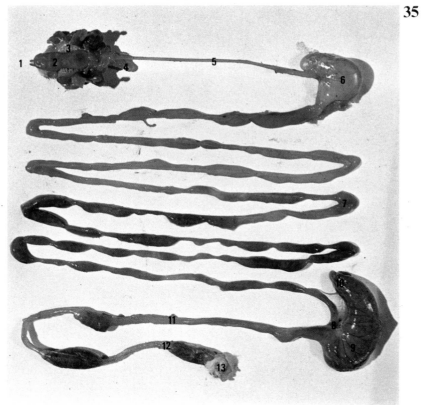

1 posterior vena cava	**10** fat body
2 left kidney	**11** left testis
3 right iliolumbar artery and vein	**12** corpus epididymis
4 seminal vesicle	**13** urethra
5 coagulating gland	**14** Cowper's gland
6 ampullary gland	**15** penis
7 urinary bladder	**16** rectum
8 prostate gland	**17** preputial gland
9 left vas deferens	**18** right testis in its scrotal sac

36 The urinogenital system of the male The alimentary tract has been removed leaving the stump of the rectum. The left testicle has been drawn out of the scrotal sac. The ventral wall (floor) of the pelvic girdle has been cut through and removed. Some fascia has been removed to display the connections of the various parts of the male genitalia. The left seminal vesicles and coagulating glands have been displaced to the right.

The seminal vesicles of the rat are large and lobulated. Each is enclosed in a capsule together with a coagulating gland.

The free communication between the peritoneal cavity and the scrotal sac is obvious on the right side.

(See also **37–39**)

Dissection procedure
Return to the carcass and examine the urinogenital system. In the male, identify the structures shown in **36**.

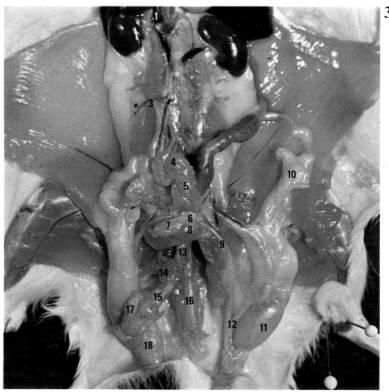

1 adrenal gland
2 left kidney
3 ureter
4 dorsal aorta and posterior vena cava
5 rectum
6 right seminal vesicle
7 right coagulating gland
8 urinary bladder
9 prostate gland
10 left vas deferens
11 fat body
12 caput epididymis
13 left testis
14 cauda epididymis
15 penis
16 preputial gland

37 A preserved preparation of the male urinogenital system The abdominal wall and other obscuring organs of the abdomen have been removed.

The spermatozoa are produced in the testis from which they first enter the epididymis, a long, fine tube which is tortuously folded on itself to produce a bulbous head (*caput*) on the anterior pole of the testis. The *caput* is linked by a narrower body (*corpus*) to the tail (*cauda*) of the epididymis.

The vas deferens is the continuation of the canal of the epididymis. Its lumen becomes less tortuous and eventually leads to the urethra, the common outlet of the male genital and urinary tracts.

The products of the seminal vesicles, and the ampullary, coagulating, prostate and Cowper's bulbourethral glands, are added to the sperm to produce the semen; they are therefore called the accessory sexual glands.

The preputial glands do not contribute to the semen. They lie just under the skin of the prepuce and discharge their contents into its cavity. They secrete musk-like compounds, the odours of which serve to identify individuals within the rat community.

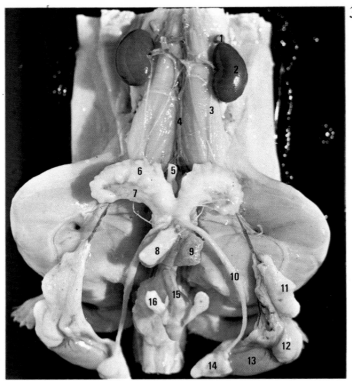

Dissection procedure
Refer to a permanent preparation of the male urinogenital system.

1 left kidney	9 vas deferens
2 ureter	10 ampullary gland
3 left spermatic artery and vein	11 seminal vesicle
4 fat body	12 coagulating gland
5 left testis	13 urinary bladder
6 caput epididymis	14 prostate gland
7 corpus epididymis	15 preputial gland
8 cauda epididymis	16 penis

38 The male urinogenital system displayed In this preserved specimen, the urinary and male genital organs have been dissected and displayed.

Large fat bodies are associated with the head of the epididymis; about half has been removed from each side of this preparation.

Dissection procedure
Refer to a permanent preparation of the male urinogenital system displayed.

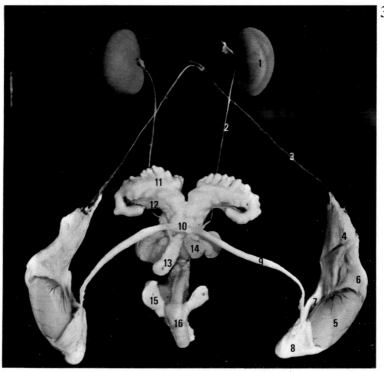

1 dorsal aorta
2 coeliac artery
3 adrenal gland
4 left adrenal artery and vein
5 anterior mesenteric artery
6 left renal artery and vein
7 left kidney
8 posterior vena cava
9 left iliolumbar artery and vein
10 right spermatic artery and vein
11 vessels of the mesentery displayed on the rat's knee
12 posterior mesenteric artery and vein
13 common iliac artery and vein
14 internal iliac artery and vein
15 external iliac artery and vein
16 rectum

17 seminal vesicle
18 pampiniform plexus
19 urinary bladder
20 prostate gland
21 left vas deferens with deferential artery and vein
22 corpus epididymis
23 left testis
24 penis
25 preputial gland
26 prepuce
27 anus
28 femoral artery and vein with saphenous nerve
29 anastomosis of spermatic and common iliac veins (left side only)
30 hepatic portal vein

39 The abdominal blood vessels of the male In this injected preparation much of the gut has been removed or reflected strongly to the right. The testes were removed from the scrotal sacs and some other components of the male genitalia displayed *in situ*. This is a detailed view of part of the specimen described in **26**.

Each spermatic vein is formed from a plexus of venules around the artery – the so-called pampiniform plexus. On the left side only there is an anastomosis between the spermatic and common iliac veins.

Dissection procedure
The injected preparation will assist in the identification of the blood vessels.

1 posterior vena cava
2 ovary contained within periovarial sac
3 fallopian tube
4 fat within the mesometrium

5 lumbar lymph node
6 rectum
7 uterus
8 urinary bladder

40 The urinogenital system of the female The abdominal portions of the alimentary tract, other than the rectum, had been removed.

The uterus consists of two horns which appear to fuse near the bladder; in fact, they retain separate openings into the vagina. Each horn terminates anteriorly in a convoluted fallopian tube which is closely applied to the ovary. The ovary is covered by a membrane which forms a periovarial sac.

This specimen contains an average amount of fat.

Dissection procedure
In the female, identify the structures shown. Examine especially the Fallopian tube, which surrounds the ovary. Note also that the rectum is situated dorsal the uterus.

1 adrenal
2 right kidney
3 renal artery and vein
4 posterior vena cava
5 dorsal aorta
6 ureter
7 ovary
8 uterus
9 mesometrium
10 urinary bladder
11 urethra
12 bisected pelvic girdle
13 vagina
14 preputial gland
15 vulva

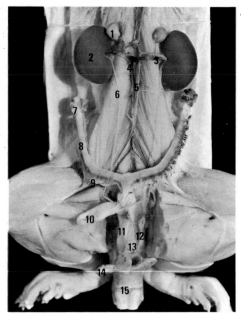

41

41 The female urinogenital system *in situ* In this preserved specimen, the abdominal wall and other overlying tissues have been removed.

A number of *corpora lutea* can be seen, especially in the left ovary.

The ova released from the ovaries are fertilised in the fallopian tubes and conveyed through them to the horns of the uterus. At parturition the foetus is expelled through the vagina. The vagina terminates in the vulva, which is that part of the female genital tract visible from the surface.

The urinary and genital tracts are quite separate in the female rat (see **8**). The female urethra, which lies ventral to the vagina, opens at the tip of the clitoris. This small homologue of the penis also has a prepuce with preputial glands like those of the male.

Dissection procedure
Refer to a permanent preparation of the female urinogenital system.

1 right renal artery and
   vein
2 kidney
3 ovary
4 ureter
5 mesometrium
6 uterus
7 rectum
8 urinary bladder
9 urethra
10 vagina
11 preputial gland
12 clitoris
13 vulva

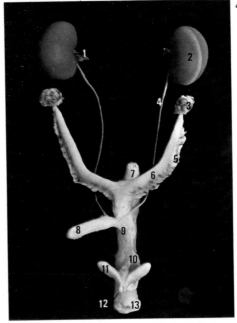

42 The female urinogenital system displayed In this preserved specimen, the urinary and female genital organs have been dissected and displayed. Each ovary contains many *corpora lutea*.

Dissection procedure
Refer to a permanent preparation of the female urinogenital system displayed.

1 adrenal
2 adrenal artery and vein
3 kidney
4 ureter
5 ovary (poorly
    developed)
6 fallopian tube
7 utero-ovarian artery
    and vein
8 iliolumbar artery and
    vein
9 ovarian artery and vein
10 uterine artery and vein
11 uterus
12 posterior vena cava
13 dorsal aorta
14 common iliac artery and
    vein
15 rectum
16 epigastric artery and
    vein
17 femoral artery and vein
    with saphenous nerve
18 pudendal artery and
    vein
19 urinary bladder
20 vesical artery and vein

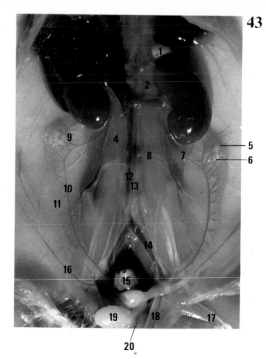

43 The abdominal blood vessels of the female This was the mal-
nourished animal with the dental abnormality shown in **14**. Although
the genitalia are immature, this dissection, which is free of abdominal
fat, shows most of the major blood vessels in the fresh state.

Note that even in this preparation the veins are easier to see than the
arteries. This difference is increased in an animal with a normal amount
of body fat since, as the arteries are usually deeper than the veins, they
are more obscured.

There are some small islands of accessory adrenal tissue near the
adrenal blood vessels; such nodules in the rat commonly contain only
cortical tissue. They could interfere with the results of experimental
adrenalectomies.

Dissection procedure
Examine the blood vessels of the female genitalia.

1 adrenal
2 perirenal fat
3–6 parts of the kidney
   3 cortex
   4 medulla
   5 pelvis
   6 pyramid
7 beginning of ureter
8 arcuate or arciform
   artery
9 ovary

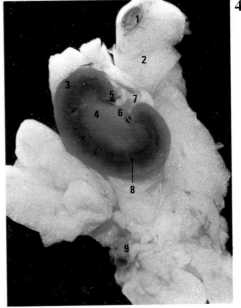

44 Longitudinal section of the kidney and the adrenal A kidney was removed from the abdominal cavity together with the adrenal gland, the ovary and portions of the perirenal fat. With a sharp knife the kidney and the adrenal were each bisected. Part of the periovarial sac was removed to expose the ovary. The tissues were floated in normal saline for photography.

The ovary contains a follicle and a number of *corpora lutea*. The cortex and medulla of the kidney are clearly distinguished. The adrenal also has two zones, a cortex and medulla, which secrete different hormones.

Dissection procedure
Remove one of the kidneys and the nearby adrenal, together with some perirenal fat. Make a longitudinal incision through the kidney and the adrenal, and examine the internal structures of these organs with a lens.

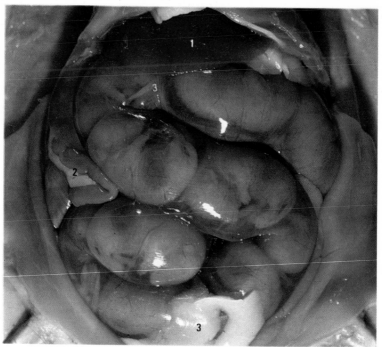

1 liver   **2** small intestine   **3** fat

45 The uterus at an advanced stage of pregnancy The appearance of the abdomen when it is first opened. Incisions have been made through the abdominal muscles and peritoneum, and the flaps of the abdominal wall pinned back.

This animal was near the end of its pregnancy ('term'). The uterus is seen to occupy the greater part of the ventral region of the abdomen.

The foetuses can be seen in their typical positions through the wall of the uterus.

Dissection procedure
A female in an advanced stage of pregnancy will appear as in **45** when the abdomen is first opened.

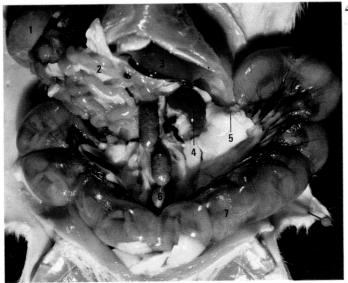

1 caecum	**5** fallopian tube
2 small intestine	**6** rectum
3 liver	**7** uterus containing eleven
4 kidney	young

46 The pregnant uterus displayed The uterus has been unfolded into the U-shape characteristic of the non-pregnant condition (see **40** or **41**), and the intestine has been displaced to the top right, except for the rectum in the mid-line which contains faecal pellets.

Note the liberal blood supply to the pregnant uterus.

There were 11 foetuses, a litter number within the normal range for the rat (the normal number of teats is 12).

Dissection procedure
Display the uterus.

1 liver
2 kidney
3 umbilical cord
4 placenta
5 amniotic sac
6 uterus

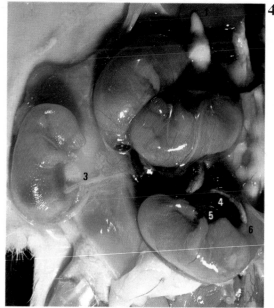

47 A further display of the foetuses, at a higher magnification.

One foetus has been removed from its membranes. The umbilical cord attaches it to the placenta. Some abdominal organs show through its thin abdominal wall.

Dissection procedure
Remove the foetuses.

1 amniotic sac
2 umbilical cord
3 placenta

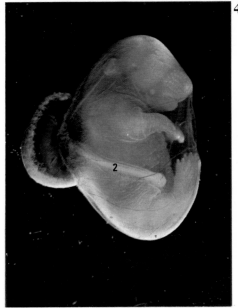

48 A foetus within its membranes This is the middle foetus on the right side of **47**; its umbilical cord sweeps across the body to the placenta.

The rows of protuberances in which are embedded the roots of the mystacial vibrissae are especially obvious. There is also a single similar protuberance behind the mouth and another just behind the eye. These are the follicles of the genal vibrissae.

Dissection procedure
Examine the foetuses along with their membranes.

1 eye
2 cerebrum
3 cerebellum
4 cervical enlargement of spinal cord
5 brachial plexus
6 scapula
7 lung
8 lumbar enlargement of spinal cord
9 right kidney
10 left kidney
11 lumbosacral plexus
12 sciatic nerve

49 The entire central nervous system *in situ* The skin, musculature and the bony roof of the cranium and vertebral column, the ribs and part of the sacrum have been removed to display the central nervous system, the spinal nerves, and some of the underlying organs. The scapulae have been reflected out into a wing-like arrangement and pinned to display the brachial nerves. This is a preserved dissection.

The spinal cord has typical cervical and lumbar enlargements.

The right kidney is situated further forward than the left; this also applies in most domestic animals, but the opposite is true in man. The kidneys are surrounded by large masses of fat.

Dissection procedure
Skin the rest of the carcass, then place it in formalin or 70 per cent alcohol to harden the nervous tissue; penetration of the fixative is helped by an incision through the skull in the mid-line over the brain. Refer to a preserved preparation of the central nervous system.

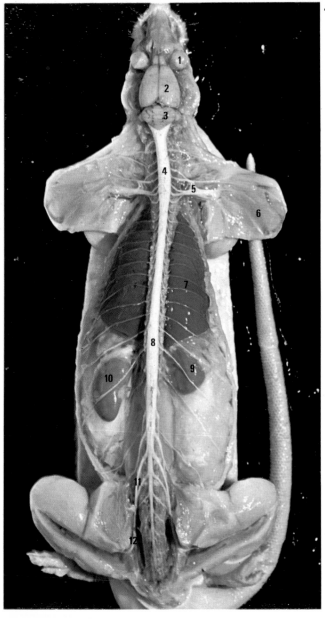

1 nasal bone
2 eye
3 muscles of eyeball
4 masseter muscle
5 olfactory lobe
6 median fissure
7 cerebral hemisphere
8 cerebellum
9 medulla oblongata
10 spinal cord
11 vertebral column
   opened to show the
   origin of the spinal
   nerves
12 brachial plexus
13 scapula, reflected

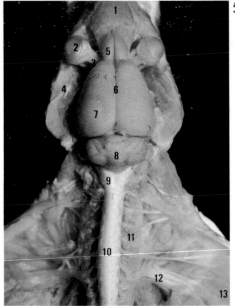

50 The brain *in situ* This is a higher magnification of the anterior end of the specimen shown in **49**.

This shows the relationship of various parts of the brain to other structures in the head and neck. Note that the muscles of the eyeball are well developed even though the rat does not move its eye. Note the great development of the masseter muscles.

The brachial plexus (see also **27**) is formed from branches of some of the cervical and thoracic spinal nerves. Such a nerve plexus serves to associate groups of muscles for combined action.

Dissection procedure
Using **49** as a guide, remove the overlying muscles and bone to expose the dorsal surface of the brain. The pineal body is usually removed with the overlying bone.

1 ophthalmic division of
   trigeminal nerve
2 olfactory lobe
3 median fissure
4 cerebral hemisphere
5 location of pineal body
   (removed)
6 posterior lobe of corpora
   quadrigemina
7 location of transverse
   sinus (removed)
8–10 parts of the
   cerebellum
     8 vermis
     9 flocculus
    10 paraflocculus
11 medulla oblongata

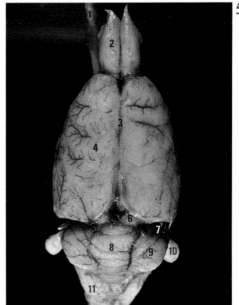

51

51 The dorsal surface of the brain The brain was fixed *in situ* by immersing the animal for three days in 70 per cent alcohol. The brain was then dissected out of the skull, and rested on a glass slide for photography. The ophthalmic division of the trigeminal nerve (see **53**) was retained on the left side, but lost on the right. The pineal body or epiphysis was also removed with the overlying bone, but the paraflocculus was retained on each side.

The rich vascular network that covers the surface of the brain is obvious. Note the relatively small flat cerebral hemispheres and the large olfactory lobes.

Dissection procedure
Continue to remove the bone from the sides of the cranium until some of the lateral features of the brain can be appreciated. Hold the preparation upside down to allow the brain to fall away from the skull as you dissect further to the ventral surface. Examine the dorsal surface of the brain thus removed.

1 incisors
2 external nasal and labial
   branches of
   trigeminal nerve
3 infraorbital fissure
4 insertion of anterior
   superficial part of
   masseter muscles
5 turbinate bones
6 major branches of
   trigeminal nerve
7 nasal septum
8 harderian gland partly
   surrounding eyeball
9 optic nerve
10 optic chiasma
11 hypophysis (pituitary)
12 tympanic bulla
13 pons
14 cochlea
15 cavity of inner ear
16 pyramid
17 roots of hypoglossal
   nerve
18 first cervical nerve
19 second cervical nerve
20 spinal cord

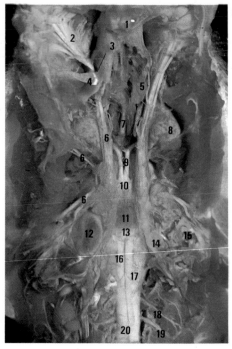

52 A stage in a dissection of the ventral surface of the brain One
photograph cannot adequately depict this area, because essential
structures such as the cranial nerves are embedded in bone, and are lost
at different stages of dissection.

The tympanic bulla houses the essential organs of hearing and of
balance. The bulla has been left intact on the right side; its bone is thin
and bulbous. On the left side it has been opened to reveal the cochlea and
other parts of the middle and inner ear.

The hypophysis (pituitary) is intact. The number and size of the
external nasal and labial termini of the trigeminal nerve emphasise the
importance for the rat of sensations from the mystacial vibrissae.

The close association of the brain with the organs of touch, smell,
vision, hearing and balance is demonstrated.

Dissection procedure
Examine the ventral surface of the brain. The distribution of the cranial
nerves, and the relationship of the ventral surface of the brain to nearby
structures, is best demonstrated by an approach through the roof of the oral
cavity on a fresh carcass.

1 ophthalmic branch of
  the trigeminal nerve (V)
2 olfactory lobe
3 olfactory tract
4 optic nerve (II)
5 cerebrum
6 maxillary-mandibular
  branch of the
  trigeminal nerve (V)
7 pituitary body or
  hypophysis
8 trigeminal nerve (V)
9 paraflocculus
10 cerebellum
11 medulla oblongata
12 spinal accessory nerve
   (XI)
13 spinal cord

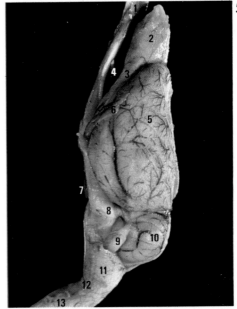

53 The brain viewed from the left side This is the same brain as that shown in **51**.

The cerebrum of the rat is relatively flat and smooth, and the ophthalmic lobes are large. The paraflocculus of the cerebellum and the hypophysis are each deeply embedded in bony cavities and are often lost in dissecting the brain; both were retained in this specimen.

Dissection procedure
Examine the lateral surface of the brain.

1 skin of muzzle	19 atlas
2 rhinarium	20 axis
3 nasal septum	21 muscles
4 hard palate	22 spinal cord
5 rugae	23 trachea
6 nasal passage	24 submaxillary salivary gland
7 ethmoid-turbinate bones	25 larynx
8 cribriform plate	26 epiglottis
9 olfactory lobe	27 opening of the eustachian tube into the nasopharynx
10 cerebral hemisphere	28 soft palate
11 lateral ventricle	29 muscles of the tongue
12 third ventricle	30 buccal cavity
13 thalamus	31 tongue
14 optic chiasma	32 mandible
15 pituitary	33 incisor
16 pineal body	34 skin folded into the diastema
17 cerebellum	
18 medulla	

54 A medial longitudinal section of the head The head was skinned, and then fixed in 70 per cent alcohol as described for **51**. It was then decalcified in 5 per cent hydrochloric acid for several days. Finally, it was cut in the mid-line with a sharp knife, and the fragments washed away.

The multiple folds of the ethmoid-turbinate bones within the nasal cavity support a large area of mucous membrane which warms the inhaled air. Olfactory cells are concentrated in the mucosa next to the cribriform plate which separates the nasal cavity from the brain. The cribriform plate has many foramina through which the nerves join the olfactory cells of the nasal mucosa to the olfactory lobes of the brain.

The eustachian (auditory or pharyngo-tympanic) tube enables air pressure to be equilibrated on either side of the tympanum; the tube has slit-like openings into the pharynx. The molars are hidden by the tongue. The space between the molars and the incisors is the diastema.

The oral cavity can be effectively sealed by the skin that is folded into

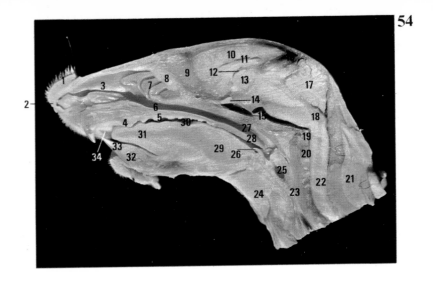

the diastema and by the apposition of the surface of the tongue to the rugose hard palate.

The slit-like lumen of the oesophagus is visible dorsal to the larynx.

Dissection procedure

Skin another carcass, and remove the head and neck. Fix them in 70 per cent alcohol or 10 per cent formalin, and decalcify in 5 per cent hydrochloric acid for three days. Cut the specimen in the mid-line with a sharp knife, wash off any fragments and identify the tissues in this median longitudinal section.

1 zygomatic arch
2 clavicle
3 gas in intestines
4 sacrum

55 An X-ray photograph of a rat, dorso-ventral and lateral views. A dead animal was used in order to avoid movement during exposure. The letter R indicates the animal's right side.

The zygomatic arch is large; it is the site of insertion of much of the rodent's well-developed masseter muscle.

The rat has a clavicle which forms a bridge between the scapula and the sternum. It is small or absent in many mammals but is well-developed in birds where it is fused over the sternum to form the familiar 'wishbone'

The fusion of the sacral vertebrae is clearly shown, as are the joint spaces of the vertebrae on either side of the sacrum.

The remarkable development of the incisors is obvious from the lateral view. The root of the lower incisor extends completely under the roots of the molars.

Dissection procedure
A better appreciation of the overall structure of the animal will be gained by study of X-ray photographs.

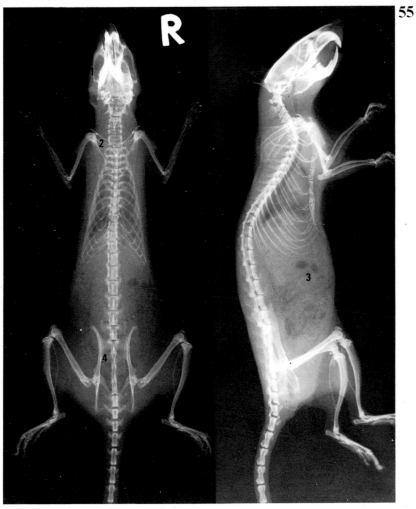

1 zygomatic arch	14 tarsus
2 mandible	15 metatarsus
3 cranium	16 phalanges
4 7 cervical vertebrae	17 os penis
5 13 thoracic vertebrae	18 scapula
6 6 lumbar vertebrae	19 humerus
7 4 sacral vertebrae	20 ribs
8 about 27 caudal vertebrae	21 sternum
9 pelvis	22 radius
10 femur	23 ulna
11 patella	24 carpus
12 tibia	25 metacarpals
13 fibula	26 phalanges

56 A transparent preparation of a whole rat, showing the skeleton stained by alizarin red. The carcass was skinned and eviscerated, and the larger muscle masses removed. It was then fixed in alcohol, defatted in acetone, cleared in sodium hydroxide, stained in alizarin red, and finally transferred to glycerol.

Alizarin stains the calcified tissues red. Apart from the bone and teeth, the cartilages of the ribs are partly calcified.

Because some musculature has been removed, the *os penis* is displaced away from the body.

The cranium of this young rat is relatively large (see **54**) and the cavities occupied by the cerebellum, cerebral hemispheres and olfactory lobes are distinguishable.

In this incompletely grown animal there is uncalcified tissue near the ends of the bones. This is the region of bone growth – in an adult animal the bones are completely calcified.

Dissection procedure

For the overall structure of the rat, study alizarin-stained transparencies.

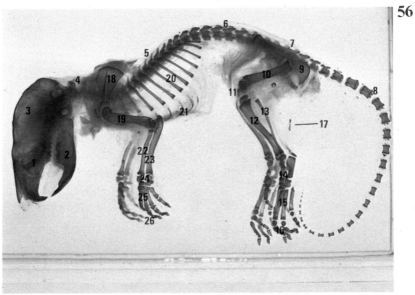

1 skull
2 zygomatic arch
3 mandible
4 tympanic bulla
5 7 cervical vertebrae
6 13 thoracic vertebrae
7 6 lumbar vertebrae
8 4 sacral vertebrae
9 about 27 caudal
   vertebrae
10 pelvis
11 femur
12 patella

13 tibia
14 fibula
15 tarsus
16 metatarsus
17 scapula
18 humerus
19 ribs
20 sternum
21 radius
22 ulna
23 carpus
24 metacarpals
25 phalanges, each of which
   terminates in a claw

57 The skeleton viewed from the left side and above This is a mounted preparation.

Note the large zygomatic arch (see **55**). The tympanic bulla is an irregularly globular bone which houses the middle and inner ear and vestibular apparatus; the external auditory opening is visible.

The grouping of the vertebrae is determined by the following (a) each thoracic vertebra bears ribs and (b) the sacral vertebrae are fused (see **55**) to form the sacrum, a part of the pelvic girdle.

Two simple clues are offered to the comparative anatomy of each limb. In the fore limb the joint between the scapula ('shoulder blade') and humerus is the shoulder, while the carpus is the wrist. In the hind limb the patella ('kneecap') sits on the knee joint and the tarsus is the ankle in man and the hock of the horse.

Dissection procedure
Study also a mounted skeleton.

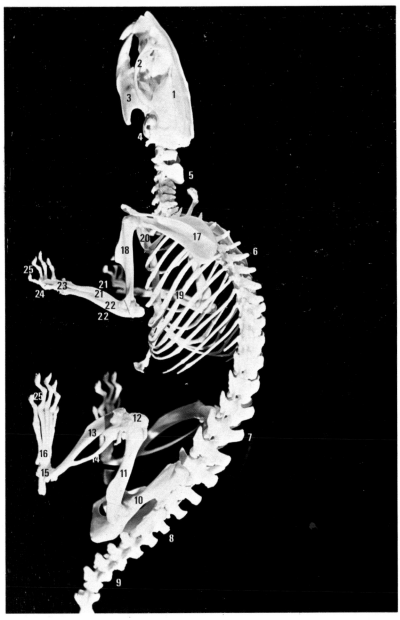

58 The brown rat, *Rattus norvegicus* The agouti colour blends readily with many natural backgrounds. The ears are relatively smaller than those of the black rat and are covered on the outer surface and part of the inner surface with short dense hairs.

The guard hairs are noticeable along the dorsum of the neck.

The eye is pigmented in the wild type (see **1**).

The vibrissae are long and enable the animal to probe widely into the environment as it advances (see **1**).

59 The black rat, *Rattus rattus* This photograph shows a number of its most characteristic features: its usual colour, its large hairless ears, its large protuberant eyes and its climbing agility – note how it uses its tail for balance on the slack rope.

General Notes
on the
Biology of the Rat

Introduction

As pests, rats of various species consume or damage ten per cent of the world's food supply. They eat and spoil grain, destroy growing crops, contaminate the food and water supply of domestic animals, kill poultry and steal eggs. In buildings, their powerful incisors cause considerable damage; gnawed gas and water pipes lead to expensive leaks and gnawed electric cables have started fires. Their gnawing and burrowing habits have caused buildings to collapse from extensive damage to the joists, partitions, walls and foundations, and roads and pavements to subside. Where rats have colonised small islands they have reached such numbers that they have almost exterminated the wild populations of rabbits and ground-nesting birds.

The brown rat, *Rattus norvegicus* belongs to the largest order of mammals, the *Rodentia*. It is one of the most numerous and cosmopolitan of mammals. Three main factors have led to its enormous success: its high rate of reproduction, its varied diet, which ensures that it seldom lacks food, and its adaptation to a variety of habitats commensal with man.

The rat's natural enemies, such as the fox, stoat, weasel, otter, owl, kestrel, buzzard and other birds of prey are gradually being reduced in numbers by man, so that periodic campaigns against the rat may be necessary.

History

Neither the brown rat (*Rattus norvegicus*) nor the black rat (*Rattus rattus*) is indigenous to Europe. They both came from the east and between them have now colonised many areas of the world. Because of the enormous amount of damage and harm they do, they are everywhere most unwelcome guests.

The black rat probably originated in India and spread through Asia Minor to north Africa and southern Europe. It may have reached western Europe during the 13th century, possibly at the time of the Crusades. Evidence for this is not clear, but it was certainly present in Great Britain long before the arrival of the brown rat.

The brown rat came to Europe early in the 18th century, possibly from China or Siberia or elsewhere north of the Himalayas, spread westwards through Russia and, within a century, had reached most European countries. It has at times been called the 'Norway rat', because it was thought to have reached Great Britain from Norway, and again the 'Hanoverian rat' because it and the equally unpopular monarch George I of the House of Hanover appeared about the same time. In most temperate regions it has replaced the black rat, except in ports and some other specialised habitats.

Habitat

Rats are found all over the world, wherever there are human settlements.

In temperate regions, the black rat is mainly confined to ports and is not found in any great numbers away from shipping, which serves as a source of periodic replenishment. Occasionally it finds a suitable environment inland, in modern blocks of centrally heated buildings. Many modern buildings have been designed to prevent colonisation by the burrowing brown rat, but the black rat, with its superior climbing power, can enter and establish itself by using telephone cables, pipes, lift shafts and plumbing systems.

The brown rat is not confined to any particular habitat. Colonies can live entirely in the wild, others occupy man-made structures including town buildings, drains and sewers, factories, mines, slaughter houses, refuse dumps, yards and farm buildings. In summer it extends into the fields, hedges and woodland of country areas. Corn ricks, providing both food and warmth, used to be a habitat of major importance to the rat, where it could continue breeding throughout the year. Changing agricultural methods, in particular, the use of the combine harvester, have now removed this niche, but the versatile rat will readily find another. Where the environment remains constant there is little movement, but occasionally the rat is forced to seek a new home, and mass migrations have been reported.

Size

The wild brown rat reaches an overall length of 37–60cm in the male, and 39–47cm in the female, the head and body being 20–37cm and 22–27cm respectively.

In general a wild rat attains a greater body weight than a laboratory rat. Its birth weight is about 5g increasing to 500g or more in the mature male but seldom beyond 350g in the female. When it is sexually mature a male weighs about 200g and a female about 115g.

External features

The brown rat has a stocky body with a blunt snout and a scaly tail, which is shorter than the combined length of the head and body. The ears are small, thick and covered with fine hairs. The eyes are black and the skin is pigmented. The soles of the feet are naked with relatively small pads; there are claws on the toes. A female usually has twelve teats, three pairs in the region of the chest and three pairs in the groin.

Its body is covered with short, soft hairs and numerous long, harsh, bristly hairs and has a shaggy appearance. The normal wild type or 'agouti grey' varies in colour, but is typically dark grey or reddish brown on the back with darker hairs along the mid-dorsal line. The belly is lighter in colour, being silvery grey or white. A dark, uniformly black type may be mistaken for a black rat, but it is uncommon.

The tail has over 210 rings of epidermal scales, between which are a few short bristles. On the head are long sensitive hairs called vibrissae, some of which attain a length greater than that of the head. They extend from the nose backwards along the upper lip, above and below the eye, and near the corner of the mouth. These hairs have sensory bulbs at their base, are well supplied with nerves and have an extensive blood supply.

Dentition

The word rodent comes from the Latin *rodere*, to gnaw. The teeth of the rat are very specialised. They are monophyodont, that is, there is only one set of teeth; there are no milk teeth as in most mammals. There are a pair of incisors, and three pairs of molars in each jaw; there are no canines and no premolars.

The gap between the incisors and the molars is called the diastema. Folds of skin from either side of the mouth can be drawn into this gap; this closes off the back of the mouth. Inedible material is not swallowed, but is rejected through the gap. This may be why chemical repellents have little effect in preventing gnawing.

The incisors are used for biting, fighting, holding food and for breaking up hard soil when excavating burrows. They are large and curved with a thick layer of orange-yellow enamel on the anterior surface. The biting edge is kept very sharp by the regular honing of the upper and lower pairs against each other, and the differential wear of the hard enamel and the softer dentine and cement on the posterior surface. They are set in very deep sockets, and have a wide open pulp cavity; they are therefore 'rootless' and continue to grow throughout life. Should one break accidentally, the opposite tooth continues to elongate at twice its normal rate to form a spiral tusk. This may eventually perforate the palate or the lip, thus causing the animal to become gagged when it will die of starvation.

The molars are ridged, with nine enamel-free cusps, arranged in three rows of three. The enamel is not pigmented. They are used for grinding food and the upper surface is abraded very slowly throughout the life of the animal.

The jaws are long and narrow. The upper row of teeth is closer together than the lower, and the grinding surfaces slope obliquely upwards and outwards. The lower jaw has a long shallow articulation; this allows both a backwards and a forwards movement which brings the incisors into apposition for gnawing, and a rotary movement which is required for the grinding action of the molars. The jaw muscles have become large and specialised to produce these movements.

The digestive system

The brown rat is omnivorous, that is, it eats a variety of both plant and animal material. It generally prefers cereals and attacks grain at all stages, when growing and when harvested. It will also eat eggs and chicks of farm-yard poultry, game birds and other ground-nesting birds, rabbits, house mice, fish, insects, spiders, offal, leather, cloth, paper and garbage of every kind. Water is essential and is drunk freely. Food found on the exploratory forays of the territory is taken back to the nest, or under cover to be eaten. Large items are carried in the incisors and smaller items in the mouth. The rat usually feeds directly off the ground but may hold the food with both hands. Soft and finely divided foods seem to be preferred to harder, coarser foods, e.g. wheatmeal to the whole wheat grains.

The muscular tongue is rough and ridged; the ridges fit into the transverse grooves on the palate. Thus in the resting position the buccal cavity is occluded, so preventing the inhalation of dust.

Food is concentrically layered within the stomach with freshly swallowed food in the centre. This retards inactivation of the salivary enzymes by hydrochloric acid.

The rat has no gall bladder; this is in general a feature of species that eat fairly continuously. The intestine is long and simple with a very large caecum.

Animal species that eat large amounts of plant material, have two methods of dealing with cellulose, which is difficult to break down. Camels and ruminants, such as the cow, have elaborate stomachs, to which regurgitated food is returned for further treatment. Rodents have a simple stomach, but additional digestion of the cellulose can take place in the caecum. To this end, and to utilise the vitamins of the B complex, which are formed, but not taken up when the food first passes through the small intestine, some rodents eat their faeces. This habit, found in rats, shrews, guinea-pigs, rabbits and hares, and probably many other animals, is known as coprophagy, or refection. It has been extensively studied in the rabbit, where two kinds of faecal pellet are formed. At night the pellet is soft and mucoid. The rabbit removes this directly from its own anus and swallows it without chewing. During the day the pellet is hard and firm and is deposited on the ground. The rat, however, forms only one kind of pellet, which is dropped on the ground. It may then pick up its own pellet, or that of another rat for re-ingestion.

The senses

The rat is a mainly nocturnal animal, displaying a circadian ryhthm, with most activity taking place at night. It uses all the senses of sight, sound, smell, taste and touch but, being nocturnal, it has very poor vision, although it will respond immediately to the slightest movement. Its senses of smell and hearing are acute, and it relies on these, rather than sight, since odours and sounds carry better at night.

The eyes of the rat are small and are placed laterally. It is capable of seeing through virtually 360, but to see something just in front, it would need to tilt its head and use one eye only. There is no advantage in moving the eyes, as there is no part of the retina which is specialised for acute vision. As the eyes are not moved, they are often capable of working independently, movement of an object on one side being discerned by one eye only. As there is a certain amount of overlap of the field of vision, the eyes can be used to judge distance.

The lens of the eye is large and almost spherical with no power of accommodation. The pupil and cornea are also large. In a small eye this means poor visual acuity. In nocturnal predators, the whole eye is enlarged, so that the image on the retina is small, but bright and clear. The retina is composed almost entirely of rods and therefore takes longer to adapt to the dark than a retina with more cones. An animal deficient in vitamin A suffers a loss of visual pigment, with decreasing powers of dark adaptation, which may lead to complete night blindness. The rat has been shown to be colour-blind.

It is well known that animals such as the cat, have eyes which appear to glow in the dark. This is often called 'eyeshine'. The eye itself does not glow but the external light is reflected back through the retina from a special layer of cells, the *tapetum lucidum*, behind the rod and cone layer. The reflected light from the tapetum increases the contrast between an object and the background, making it more discernible in the dark. Some species with 'eyeshine' do not have a definite tapetum. Only one rodent, *Cuniculus paca* is known to possess a tapetum; the structural basis of the rat's 'eyeshine' is unknown. The 'eyeshine' of rats appears red, whereas that of cats and other predators is bright green.

Smaller mammals, in general, have long narrow basilar membranes in the ear, which are suitable for both high and low frequency hearing. Some correlation between the size and weight of an animal and its hearing range has been found. The voice of the rat is high-pitched, and the range of notes it can appreciate is different from that of man. At 8 kilohertz man and rat are about equally sensitive, below this level the rat is poorer than man, at higher frequencies the rat is more sensitive. The rat can also make and hear sounds above 20 kilohertz, which is the limit of human hearing and, therefore, in the ultrasonic frequency range for man. The ultrasounds are probably produced by a whistle-like mechanism, associated with respiration. At the onset of exhalation the sound is emitted, normally through the mouth. The nose and mouth are not used as a resonance cavity, and the exact structures in the larynx with which the sound is made have still to be identified.

Baby rats use ultrasounds in conditions of 'stress', when cold or hungry. At about three weeks of age, when their dependence on the mother ends, the distress calls also finish. Adult rats produce short and long pulses of ultrasound. The short pulses seem to be aggressive, and the long pulses submissive. Rats may be capable of finding their way through a maze by using sonic waves, from such activities as sniffing, sneezing and scratching the floor, but not from the high frequencies used by the bat.

The olfactory lobes and hippocampi of the brain of the rat are large, indicating the importance of smell. All food and all strange objects, as well as other rats are sniffed; a rat emerging from its burrow uses its sense of smell rather than its sight to detect danger. Urine is one of the least specialised conveyers of odour. The preputial glands introduce their secretion into the urine via a duct which opens near the urethral orifice. The dermal sebaceous glands are the most commonly used exudative organs for odour production. In some mammals they are concentrated in thickened glandular complexes but in the rat they are widely scattered, except on the feet, where there are densely packed sebaceous glands.

The rat uses its own odour to mark the home territory and thus advertise occupancy of a particular area. The establishment and maintenance of a suitable tract of land is largely the rôle of the male. Although obvious scent glands and clearly defined marking behaviour is lacking, there seems to be some relationship between the dominance status of the marker and the stimulus the odour gives. Odour also helps the rat to discriminate between its own and other species and between separate populations, thus giving group cohesion. In reproduction, odour brings about successful mating or prevents mating when both partners are not in breeding condition. There seems to be an odour bond between mother and offspring. A special locomotor-inhibiting substance is produced by the female when the young are 14 days' old and ceases on weaning at 27 days' old. The site of production of the substance is not known and it is not family specific.

The long vibrissae of the rat are extremely sensitive organs of touch. They are particularly well supplied with blood vessels and nerves and, probably, together with a keen sense of smell, partially compensate for poor vision. They are an aid in locomotion, for they help the animal determine the nearness and position of edges and corners, discriminate inequalities of the surface, and keep its equilibrium. When moving along a narrow support the vibrissae are turned down and trailed along the edge. Rats disturbed in the open quickly find the safety of a wall. The vibrissae brush the sides of the wall and with this contact, the rat travels quickly and confidently. When experimental rats have to gauge depths within reach of their vibrissae, tactile sensations override visual ones.

Breeding

The breeding season is mainly in the spring and autumn but continues all year where conditions are favourable. If the young are at risk from cold weather, breeding may cease in the winter; even in a continuous cycle there are peaks of pregnancy rates, the main one in Great Britain being from March to June. The female usually has six to ten young, with an average of eight per litter. The young become sexually mature at three to six months.

A pregnant female builds a substantial nest to keep her young warm, as they are unable to control their body temperature for the first few days of life. Gestation takes 20–24 days. During birth the mother stands, and helps to free the young from the membranes. She eats the membranes, the umbilical cord and the placenta; occasionally she may even eat the young. Parturition usually takes 1–2 hours for the entire litter. The new-born young are pink-skinned, hairless and quite helpless: they are blind, have closed ear passages, no teeth and are unable to walk. The mother suckles her litter for three weeks. Body hair appears about the 4th day, the ears are open about the 11th, and the eyes about the 14th day. During their period of helplessness the mother will retrieve her young when they stray from the nest, move them to a safer site when danger threatens, and defend them fiercely from predators and other rats. The male takes no direct part in caring for them. By the 4th week the young have teeth, can hear, see, run about and feed on a mixed diet and, therefore, are capable of looking after themselves.

Even before her litter is weaned, the female may be pregnant again. The rat is polyoestrous, with a short interval of four days between successive periods of ovulation. She has, therefore, a very high breeding potential. Although the average number per litter is eight, the total number of progeny varies with the weight of the female. The heavier she is, the more she has in each litter, as the number of ova liberated is related to body size, and the more pregnancies she has. The average number ovulated at once is ten, which is high for a placental mammal. However, there are a number of intra-uterine deaths; 60 per cent of litters will have some losses before reaching full-term, although few litters are lost entirely.

The great fecundity of the rat enables it to replace very quickly those killed in natural disasters and in inefficient control campaigns, and has undoubtedly contributed to the enormous success of the species.

Behaviour

Social behaviour

Rats are social animals. They form a family group of mother and young. Adults live together and sleep piled on top of each other, not necessarily for warmth, as they do this in heat-controlled laboratories. Mutual cutaneous stimulation seems to be important and is provided in a number of ways: they huddle together, groom each other, crawl under and over another member of the group, and nose the flank of other rats. This tactile stimulation is very necessary when the young are born.

Experimentally preventing the mother from licking her offspring leads to their death; the reflexes needed for defaecation and urination seem to need cutaneous stimulation of the genito-anal region.

Their social behaviour for the most part consists of stereotyped responses to signals, such as odours, sounds and postures. The preputial glands secrete musk-like compounds; the odours of these substances play a significant part in the recognition of another member of the colony, of a stranger, or of a female in oestrous. Sound is also important in behaviour. Young rats squeak when they have strayed from the nest and thus elicit retrieval by their mother. Adult rats make whistling noises, which may prevent their being attacked. The scream of a cornered rat acts as an alarm to other rats and may deter the predator. Two postures, which the rat may adopt when approached by another, could have social significance. The animal may appear 'submissive': it lies on one side with closed eyes. In this position it is usually not attacked. The other is a 'threatening' posture: one side is turned towards the intruder, the back is arched and the legs are fully extended. This may promote withdrawl of the intruder.

Territory

Rats live in colonies, which may number many hundreds, and have common nesting sites and feeding grounds. The home range of each rat is quite small and is defended against other members of their species by the male; the female defends her nest only when it contains young. The larger area occupied by the colony is defended by the males against outside intruders. Rats spend most of the day in the nest, resting and sleeping. In the evening and at night they travel along definite paths to the feeding ground. These pathways are often near or under cover and in built-up areas they usually stay close to walls. Use of regular routes leaves odour trails, which other rats follow, and on hard surfaces the runs are marked in places by black, greasy, odorous smears.

Within the colony there is no 'pecking order' or 'linear hierarchy'. Males can live together without much conflict, although there is some internal strife. There are a number of dominant males to which all the others are subordinate. It appears that any male which cannot live within this system dies. Only adult males from another colony will be attacked. The defending rat normally displays a stereotyped behaviour pattern: it chatters its teeth, urinates, defaecates, raises its hair, adopts a threatening posture, leaps and bites. It may also peacefully groom its opponent! The sequence and intensity of these components is variable, but usually the outcome is the withdrawal of the intruder without much harm to either rat. However, even though the intruder is not wounded, it may sicken and die in a day or two if it cannot escape; this situation may occur, for example, in caged animals.

Nests and burrows

The nest of the brown rat provides a place of concealment and warmth during the day. It is made from any available material – rags, straw, wood shavings, sacking, paper and sticks – which is carried in the mouth. In warm conditions the nest is loosely constructed. The colder the weather, the better it is made, but the pregnant female builds a substantial nest, irrespective of the temperature.

The nest is often sited in a burrow, part of which is enlarged to form a 'den'. A burrow is made wherever the soil is suitable for digging, but especially in river banks and hedgerows, and under the foundations of buildings such as chicken houses. It may have a number of exits, some of which are concealed, and blind branches in which faeces are deposited. Food is taken to the nest to be eaten in safety, out of sight of enemies. Food and useless objects may be stored in the branches of the burrow.

The rat uses its front feet to scratch out the tunnels; it pushes the loosened soil under its chest and belly and finally kicks it backwards with its hind feet. The heap of soil is then pushed along the burrow to the entrance with the front feet and head and scattered in a crescentic shape to the front and sides. Its teeth are used to remove obstructing roots and to break up compacted soil.

Control

Wild rats are very difficult to eradicate, as they are markedly neophobic; that is to say, any strange object, harmless or otherwise, in their familiar environment will be avoided. When curiosity overrides their fear, only small amounts of strange food will be taken. This neophobia is essential in an enterprising and experimental feeder (as contrasted with a strictly traditional feeder like the guinea-pig). It is the only way of finding out whether a strange object is edible or harmful.

The two main types of rodenticide used are acute and chronic poisons. The rat feels the ill effects of acute poisons so rapidly that if it has eaten a sub-lethal dose, it will avoid any food containing this substance in the future. Pre-baiting is used to ensure that the rat eats a sufficient quantity of the poisoned bait. To overcome the rat's neophobia, harmless food is put down for a number of nights; when it is no longer bait-shy, the acute poison is added. Some of these poisons are toxic to man and his domestic animals and, as they may have no effective antidotes, they are hazardous to use.

The chronic rodenticides are mainly anticoagulants. Their effect is cumulative; a single dose is seldom lethal. They do not require pre-baiting. Instead, the poisoned bait is put down for a number of days or weeks which gives the rat the opportunity to eat a sufficient quantity. The rat does not feel ill until it has taken a lethal dose and so it does not become bait-shy. These chronic poisons are much safer to use, since domestic animals are less likely to take repeated doses, and their effects are reversible.

The first anticoagulant to be used against rats in the 1940s was a derivative of dicoumarol. Naturally-occurring dicoumarol had been found to cause 'sweet clover disease', a haemorrhagic disease of cattle, which was widespread in the United States in the 1930s. The dicoumarol derivative was used extensively for a number of years but, in 1960, brown rats near Glasgow were found to be resistant to it. Since then, pockets of inherited resistance have arisen elsewhere. Other anti-coagulants have been tried on the genetically-resistant strains but, unfortunately, the rats may also become resistant to these.

Other methods of control can be used where there are anticoagulant resistant rats, or other difficulties to overcome. Trapping and gassing are particularly useful in the countryside. Fumigation is sometimes the best method in food stores, or where there are stocks of grain or flour, such as in the holds of ships. Possibly the best and easiest way to control the rat population once a poison campaign has drastically reduced the numbers, is to improve the hygiene of the infested area. This should remove the food source, the shelter and the nesting material necessary for the colony to thrive.

Diseases

Zinsser (1935) suggests that rat-borne diseases have shaped the history of mankind, since some of them are readily transmitted to man or to his domestic animals. Because of its close association with human habitations, there are greater opportunities for the transmission of disease from the rat than from other wild animals.

Bubonic plague is caused by the bacterium *Yersinia pestis*. It is spread by the flea, *Xenopsylla cheopis*, which attacks man when its usual host, the black rat, dies. It can also be carried by a flea found chiefly on the brown rat. Plague is still endemic in rats and other rodents in certain countries. Although epidemics of plague have ravaged mankind since pre-Christian times, in recent centuries they have been relatively minor and infrequent. One of the most notorious of many plague epidemics was the Black Death of medieval times. The last major epidemic in Great Britain was the Great Plague of London in 1665.

Rat-bite fever occurs sporadically in most countries. Two distinct diseases may result when an infected rat bites man. One is caused by *Streptobacillus moniliformis* and the other, called Sodoku, by *Spirillum minus*. These two bacteria are commonly found in the mouth of the rat.

The kidneys of the common rat are the most important reservoir of certain species of *Leptospira*. These spirochaetes are excreted in the urine of the rat, throughout most of its life. The older the rat, the more likely it is to be a carrier. The spirochaete can penetrate the human skin through small abrasions. It enters the blood stream and causes leptospirosis or Weil's disease, an occupational hazard of people who work in wet, rat-infested conditions. It is a particular risk to farm workers.

By eating incompletely cooked pork man could acquire the parasitic nematode *Trichinella spiralis*, which the pig has obtained by ingesting the muscle of a rat. The common tapeworm of rats *Hymenolepis nana* can also affect man.

Other diseases for which various species of rats are important sources of human infection are salmonellosis, scrub typhus, murine typhus and Lassa fever. Leptospirosis and salmonellosis are also frequently transmitted to domestic animals.

There are many other rat infections that are potentially transmissible to man or his animals (Mesina and Campbell, 1975), but opportunities for their transmission appear to be more limited.

No infectious disease has been successfully exploited in the control of wild rats, as myxomatosis has in rabbits, nor does there appear to be any immediate prospect of such a development.

Of course, the rat's non-transmissible diseases are also of great interest. Many diseases may interfere with the conduct of laboratory experiments and, therefore, have been extensively studied by the laboratory animal scientist (Tuffery and Innes, 1963; Cotchin and Roe, 1967). Some are of particular interest to the comparative pathologist, since they serve as models of diseases in man and other animals. Much of our knowledge of nutritional diseases, especially those resulting from vitamin deficiencies, is based on experimental studies on the rat.

The laboratory rat

Second only to the mouse, the rat is probably the most commonly used laboratory animal. The wild rat is a dangerous experimental animal and may attack a handler, but the domesticated brown rat takes readily to laboratory conditions. The albino, which has been developed by selection over the last seventy years from *Rattus norvegicus*, differs in many important aspects of behaviour from its wild progenitor.

It has been selected for tameness and differs from the wild rat in its behaviour towards man. There is a reduced tendency to attack smaller animals, to flee from man, or to struggle or bite on being handled. However, if it is badly handled or suffers from certain nutritional deficiencies, it can become savage. Fighting between adult males is less ferocious, and fear of new objects has been greatly modified. Its choice of food may also differ from that of a wild rat. Physically, the laboratory rat attains a lower body weight, and it has smaller adrenals and a smaller brain and spinal cord. It is also less resistant to cold as it seems to lack the ability to grow a thicker coat in cold conditions.

The rat is used for an enormously wide variety of purposes, including experimental pathology, biological assay, toxicity studies, nutritional research, cancer research and teaching. The rat is a traditional animal for behavioural research. Its ability to find its way to and through the branching passages of the burrow, enable it to find its way readily through a maze. Similarly, its ability to handle food and to build nests is basic to the quite complicated manipulations required in other behavioural studies. There is a growing use of gnotobiotic rats in various research fields – these are germ-free animals which have been deliberately exposed to a known microbial flora. Germ-free or axenic rats have no microbial flora and are obtained by a special aseptic Caesarian operation on the full-term dam. Gnotobiotic guinea-pigs, pigs, dogs, cats and chickens are also used, but not as frequently as rats and mice. They are expensive to produce and maintain but for some experiments they are essential.

There are many colonies of rats in existence, but the best known of the albino rats is that bred at the Wistar Institute of Anatomy and Biology at Philadelphia, U.S.A. Apart from the familiar albino, with pink eyes and poor sight, there are various other strains of laboratory rats, some hooded, piebald, grey or black, many of which have pigmented eyes and, therefore, better vision.

Under the controlled conditions of the laboratory there is no particular breeding season. Breeding begins at about eleven weeks of age

INDEX

Figures in light type refer to page numbers
Figures in **bold** type refer to picture numbers

Abdomen, 14
Abdominal wall, **12**, 20
Adrenal, **20, 21, 22, 41, 43, 44**, 32, 33, 67, 102
Agouti, **50**, 86, 92
Albino, **1**, 9, 102
Alimentary tract, **19, 53**, 30, 52
Anticoagulants, 100
Anus, **7, 8, 10, 11, 32, 35, 39**, 16, 17, 19, 30, 46, 52, 60
Aorta, **21, 22, 24, 25, 27, 30, 37, 39, 41, 43**, 33, 34, 36, 37, 40, 43, 56, 60, 64, 66
   arch of, **26, 27, 38**, 40
Artery
   adrenal, **26, 39, 43**, 38, 60, 66
   arciform, **44**, 67
   arcuate, **44**, 67
   axillary, **27**, 40
   brachial, **26**, 38
   carotid
      common, **15, 26, 27**, 25, 38, 40, 43
      external, **15**, 25
      internal, **15**, 25
   coeliac, **22, 39**, 34, 60
   deferential, **39**, 60
   epigastric, **43**, 66
   femoral, **26, 39, 43**, 38, 60, 66
   iliac
      common, **26, 39, 43**, 38, 60, 66
      external, **26, 39**, 38, 60
      internal, **39**, 60
   iliolumbar, **26, 36, 39, 43**, 38, 54, 60, 66
   innominate, **26**, 38, 43
   mesenteric
      anterior, **21, 22, 26, 39**, 33, 34, 38, 60
      posterior, **21, 26, 39**, 33, 38, 60
   ovarian, **43**, 66
   pudendal, **43**, 66
   pulmonary, **30**, 43
   renal, **26, 39, 41, 42**, 38, 60, 64, 65
   spermatic, **38, 39**, 58, 60
      internal, **26**, 38
   subclavian, **26**, 38, 43
   submental, **15**, 25
   superficial temporal, **16**, 26
   thoracic, lateral, **26**, 38
   uterine, **43**, 66
   utero-ovarian, **43**, 66
   vesical, **43**, 66
Appendix, **32, 35**, 46, 52
Atlas, **54**, 78
Auditory tube, 79

Auricle
   left, **25**, 37
   right, **24, 27**, 36, 40
Axenic rats, 102
Axis, **54**, 78

Bait-shyness, 99, 100
Behaviour, 97–100
Bile duct, **32, 33**, 46, 48
Birth, 97
Black rat, **59**, 87, 90, 91
Bladder
   gall, 48, 93
   urinary, **18, 19, 36, 37, 38, 39, 40, 41, 42, 43**, 28, 30, 54, 56, 58, 60, 62, 64, 66, 74
Brain
   cerebellum, **49, 50, 51, 53, 54**, 72, 74, 75, 77, 78, 83
   cerebral hemispheres, **50, 51, 54**, 74, 75, 78, 83
   cerebrum, **49, 53**, 72, 77
   corpora quadrigemina, **51**, 75
   dorsal surface, 75
   flocculus, **51**, 75
   hippocampi, 95
   infraorbital fissure, **52**, 76
   median fissure, **50, 51**, 74, 75
   medulla, brain, **54**, 78
      oblongata, **50, 51, 53**, 74, 75, 77
   olfactory lobes, **50, 51, 53, 54**, 74, 75, 77, 78, 82, 95
   ophthalmic lobes, **53**, 77
   optic chiasma, **52, 54**, 76, 78
   paraflocculus, **51, 53**, 75, 77
   pons, **52**, 76
   thalamus, **54**, 78
   ventral surface, 76
   ventricle, **54**, 78
   vermis, **51**, 75
Breeding, 96, 102
Bristles, 11
Bronchus, **30**, 43
Brown rat, **58**, 86, 90, 91, 92
Buccal cavity, **54**, 24, 44, 78, 93
Burrows, 99

Caecum, **12, 18, 19, 22, 32, 35, 46**, 20, 28, 30, 34, 46, 52, 69, 93, 94
   apex of, **34**, 50
Caput epididymis, **37, 38**, 56, 58
Carpus, **56, 57**, 82, 84
Cartilage
   costal, 36
   thyroid, **28, 29**, 41, 42
   xiphoid, **12, 19, 23**, 20, 30, 35

Central nervous system, **49**, 72
Clavicle, **55**, 80
Claw, **4**, **5**, **57**, 12, 13, 84
Clitoris, **6**, **8**, **42**, 14, 17, 64, 65
Cochlea, **52**, 76
Coeliac axis, **21**, 33
Cold, resistance to, 102
Colon, **18**, **19**, **32**, **34**, **35**, 28, 30, 46, 50, 52
Colonies, 98
Colour-blindness, 95
Control, 99, 100
Coprophagy, 94
Corpora lutea, 64, 65, 67
Cranium, **56**, 82
Cribriform plate, **54**, 78
Cutaneous stimulation, 97

Damage caused by rats, 90
Defaecation reflex, 98
Dental formula, 10
Dentition, 92
Determination of sex, 17, 19
Diaphragm, **23**, **24**, **25**, **26**, 35, 36, 37, 38
Diastema, **2**, **31**, **54**, 10, 44, 78, 92
Dicoumarol, 100
Diet, 93
Differences between brown & black rat, 104, 105
Digestive system, 93, 94
Digit/digital pad, **4**, **5**, 12, 13
Disease, rat-borne, 100, 101
Duct
    bile, **32**, **33**, 46, 48
    exorbital lachrymal gland, **16**, 26
    parotid, **14**, **16**, 24, 26
Duodenum, **32**, **33**, **34**, 46, 48, 50

Ear, 19, 84, 86, 87, 92, 95, 97, 105
Enemies of the rat, 90
Epidemics, 101
Epididymis
    caput, **37**, **38**, 56, 58
    cauda, **37**, **38**, 56, 58
    corpus, **36**, **38**, **39**, 54, 56, 58, 60
Epiglottis, **28**, **29**, **54**, 41, 42, 78
Ethmoid turbinate bones, **54**, 78
Eustachian tube, **54**, 78
External features of brown rat, 92, 105
External genitalia, female, **8**, 17
    male, **7**, **9**, 16, 18
    orifice of the urethra, 18
Eye, **16**, **49**, **50**, 19, 26, 72, 74, 86, 87, 92, 94, 95, 97, 105
'Eyeshine', 95

Faecal pellets, 30, 46, 50, 69, 94
Faeces, 94, 99

Falciform ligament, **23**, 35
Fallopian tube, **40**, **43**, **46**, 62, 66, 69
Fecundity of the rat, 97
Feeding grounds, 98, 105
Feet, 92, 99, 105
Femur, **56**, **57**, 82, 84
Fibula, **56**, **57**, 82, 84
Foetus, 68, 70, 71
Food, 99
Foot, **4**, **5**, 12, 13
Fore limb, 9, 40

Gait, 12
Gall bladder, 48, 93
Genital aperture, **8**, 17
    papilla, **10**, **11**, 19
    tract, 18, 64
Germ-free rats, 102
Gestation period, 97
Gland
    abdominal, 22
    accessory sexual, **37**, 56
    adrenal, **20**, **21**, **37**, **39**, 32, 33, 56, 60
    ampullary, **36**, **38**, 54, 56, 58
    coagulating, **36**, **37**, **38**, 54, 56, 58
    Cowper's, **36**, 54, 56
    harderian, **16**, **52**, 26, 76
    hibernating, 25
    inguinal, 22
    lachrymal, **12**, 20
        exorbital, **14**, **16**, 24, 26
        intraorbital, **16**, 26
    preputial, **13**, **18**, **19**, **36**, **37**, **38**, **39**, **41**, **42**, 22, 28, 30, 54, 56, 58, 60, 64, 65, 95, 98
    prostate, **18**, **36**, **37**, **38**, **39**, 28, 54, 56, 58, 60
    salivary, **12**, 20, 24
        parotid, **14**, **16**, 24, 26
        sublingual, **14**, 24
        submaxillary, **14**, **54**, 24, 78
    scent, 96
    sebaceous, 95, 96
Gnotobiotic rats, 102

Habitat, 91, 104
Hair, 9, 86, 92, 97
'Hanoverian' rat, 91
Head, section of, **54**, 79
    superficial tissues of, **16**, 26
Hearing range, 95
Heart, **23**, **25**, **26**, **28**, **29**, **32**, 35, 37, 38, 41, 42, 46
History, 90
Home range, 98
Humerus, **56**, **57**, 82, 84
Hypophysis, **52**, **53**, 76, 77